From the Closed World

TO THE

INFINITE

UNIVERSE

Publications of the

Institute of the

History of Medicine,

THE JOHNS HOPKINS UNIVERSITY

Third Series: THE HIDEYO NOGUCHI LECTURES,

Volume VII

Alexandre Koyré

From the Closed World TO THE INFINITE UNIVERSE

ALEXANDRE KOYRÉ

The
Johns
Hopkins
University
Press / BALTIMORE AND LONDON

Printed in the United States of America on acid-free paper

Johns Hopkins Paperbacks edition, 1968
06 05 04 03 02 01 00 99 98 97 13 12 11 10

The Johns Hopkins University Press
2715 North Charles Street
Baltimore, Maryland 21218-4319
The Johns Hopkins Press Ltd., London

Library of Congress Catalog Card Number 57-7080

ISBN 0-8018-0347-0

A catalog record for this book is available from the British Library.

The Hideyo Noguchi Lectureship

In 1929 the late Dr. Emanuel Libman of New York gave $10,000 to The Johns Hopkins University for the establishment of a lectureship in the History of Medicine. In accordance with Dr. Libman's wishes it was named The Hideyo Noguchi Lectureship to pay tribute to the memory of the distinguished Japanese scientist.

The present volume owes its origin to the eleventh lecture on this foundation which was delivered on December 15, 1953, at The Johns Hopkins Institute of the History of Medicine by Professor Alexandre Koyré.

Preface

Time and again, when studying the history of scientific and philosophical thought in the sixteenth and the seventeenth centuries — they are, indeed, so closely interrelated and linked together that, separated, they become ununderstandable — I have been forced to recognize, as many others have before me, that during this period human, or at least European, minds underwent a deep revolution which changed the very framework and patterns of our thinking and of which modern science and modern philosophy are, at the same time, the root and the fruit.

This revolution or, as it has been called, this " crisis of European consciousness," has been described and explained in many different ways. Thus, whereas it is generally admitted that the development of the new cosmology, which replaced the geo- or even anthropocentric world of Greek and medieval astronomy by the heliocentric, and, later, by the centerless universe of modern astronomy, played a paramount role in this process, some historians, interested chiefly in the social implications of spiritual changes, have stressed the alleged conversion of the human mind from *theoria* to *praxis*, from the *scientia contemplativa* to the *scientia activa et operativa*, which transformed man from a spectator into an owner and master of nature; some others have stressed the replacement of the teleological and organismic pattern of thinking and explanation by the mechanical and causal pattern, leading, ultimately, to the " mechanisation of the world-view " so prominent in modern times, especially in the eighteenth century: still others have simply described the despair and confusion brought by the " new philosophy " into a world from which all coherence was gone and in which the skies no longer announced the glory of God.

As for myself, I have endeavored in my *Galilean Studies* to define the structural patterns of the old and the new world-views and to determine the changes brought forth by the revolution of the seventeenth century. They seemed to me to be reducible to two fundamental and closely connected actions that I characterised as the destruction of the cosmos and the geometrization of space, that is, the substitution for the conception of the world as a finite and well-ordered whole, in which the spatial structure embodied a hierarchy of perfection and value, that of an indefinite or even infinite universe no longer united by natural subordination, but unified only by the identity of its ultimate and basic components and laws; and the replacement of the Aristotelian conception of space — a differentiated set of innerworldly places — by that of Euclidean geometry — an essentially infinite and homogenous extension — from now on considered as identical with the real space of the world. The spiritual change that I describe did not occur, of course, in a sudden mutation. Revolutions, too, need time for their accomplishment; revolutions, too, have a history. Thus the heavenly spheres that encompassed the world and held it together did not disappear at once in a mighty explosion; the world-bubble grew and swelled before bursting and merging with the space that surrounded it.

The path which led from the closed world of the ancients to the open one of the moderns was, as a matter of fact, not very long: barely a hundred years separate the *De revolutionibus orbium coelestium* of Copernicus (1543) from the *Principia philosophiae* of Descartes (1644); barely forty years these *Principiae* from the *Philosophia naturalis principia mathematica* (1687). On the other hand, it was rather difficult, full of obstacles and dangerous road blocks. Or, to put it in simpler language, the problems involved in the infinitization of the universe are too deep, the implications of the solutions too far-reaching and too important to allow an unimpeded progress. Science, philosophy, even theology, are, all of them, legitimately interested in questions about the nature of space, structure of matter, patterns of action and, last but not least, about the nature, structure, and value of human thinking and of human science. Thus it is science, philosophy, and theology, represented as often as not by the very same men — Kepler and Newton, Descartes and Leibniz — that join and take part in the great debate that starts with Bruno and Kepler and ends — provisionally, to be sure — with Newton and Leibniz.

I did not deal with these problems in my *Galilean Studies*, where I had to describe only the steps that led to the great revolution and formed, so to speak, its prehistory. But in my lectures at The Johns Hopkins University — "The Origins of Modern Science," in 1951, and "Science and Philosophy in the Age of Newton," in 1952 — in which I studied the history of this revolution itself, I had the opportunity to treat as they deserved the questions that were paramount in the minds of its great protagonists. It is this history that, under the title *From the Closed World to the Infinite Universe*, I have endeavored to tell in the Noguchi Lecture that I had the honour of giving in 1953; and it is the self-same story that, taking the history of cosmology, as Ariadne's thread I am retelling in this volume: it is, indeed, only an expanded version of my Noguchi Lecture.

I would like to express my gratitude to the Noguchi Committee for its kind permission to expand my lecture to its present dimensions, and to thank Mrs. Jean Jacquot, Mrs. Janet Koudelka, and Mrs. Willard King for assistance in preparing the manuscript.

I am also indebted to Abelard-Schuman, publishers, for the permission to quote Mrs. Dorothea Waley Singer's translation of Giordano Bruno's *De l'infinito universo et mondi* (New York, 1950).

Alexandre Koyré

PRINCETON

Contents

ILLUSTRATIONS

Introduction

It is generally admitted that the seventeenth century underwent, and accomplished, a very radical spiritual revolution of which modern science is at the same time the root and the fruit.[1] This revolution can be — and was — described in a number of different ways. Thus, for instance, some historians have seen its most characteristic feature in the secularization of consciousness, its turning away from transcendent goals to immanent aims, that is, in the replacement of the concern for the other world and the other life by preoccupation with this life and this world. Some others have seen it in the discovery, by man's consciousness, of its essential subjectivity and, therefore, in the substitution of the subjectivism of the moderns for the objectivism of mediaevals and ancients; still others, in the change of relationship between θεωρία and πρᾶξις, the old ideal of the *vita contemplativa* yielding its place to that of the *vita activa.* Whereas mediaeval and ancient man aimed at the pure contemplation of nature and of being, the modern one wants domination and mastery.

These characterizations are by no means false, and they certainly point out some rather important aspects of the spiritual revolution — or crisis — of the seventeenth century, aspects that are exemplified and revealed to us, for example, by Montaigne, by Bacon, by Descartes, or by the general spread of skepticism and free thinking.

Yet, in my opinion they are concomitants and expressions of a deeper and more fundamental process as the result of which man — as it is sometimes said — lost his place in the world, or, more correctly perhaps, lost the very world in which he was living and about which he was thinking, and had to transform and replace not only his fundamental concepts and attributes, but even the very framework of his thought.

This scientific and philosophical revolution — it is indeed impossible to separate the philosophical from the purely scientific aspects of this process: they are interdependent and closely linked together — can be described roughly as bringing forth the destruction of the Cosmos, that is, the disappearance, from philosophically and scientifically valid concepts, of the conception of the world as a finite, closed, and hierarchically ordered whole (a whole in which the hierarchy of value determined the hierarchy and structure of being, rising from the dark, heavy and imperfect earth to the higher and higher perfection of the stars and heavenly spheres),[2] and its replacement by an indefinite and even infinite universe which is bound together by the identity of its fundamental components and laws, and in which all these components are placed on the same level of being. This, in turn, implies the discarding by scientific thought of all considerations based upon value-concepts, such as perfection, harmony, meaning and aim, and finally the utter devalorization of being, the divorce of the world of value and the world of facts.

It is this aspect of the seventeenth century revolution, the story of the destruction of the Cosmos and the infinitization of the universe that I will attempt to present here, at least in its main line of development.[3]

The full and complete history of this process would make, indeed, a long, involved and complicated story. It would have to deal with the history of the new astronomy in its shift from geocentrical to heliocentrical conceptions and in its technical development from Copernicus to Newton, and with that of the new physics in its consistent trend toward the mathematization of nature and its concomitant and convergent emphasis upon experiment and theory. It would have to treat the revival of old, and the birth of new, philosophical doctrines allied with, and opposed to, the new science and new cosmological outlook. It would have to give an account of the formation of the "corpuscular philosophy," that strange alliance of Democritus and Plato, and of the struggle between the "plenists" and the "vacuists" as well as that of the partisans and the foes of strict mechanism and attraction. It would have to discuss the views and the work of Bacon and Hobbes, Pascal and Gassendi, Tycho Brahe and Huygens, Boyle and Guericke, and of a great many others as well.

However, in spite of this tremendous number of elements, discoveries, theories and polemics that, in their interconnections, form the complex and moving background and sequel of the great revolution, the main line of the great debate, the main steps on the road which leads from the closed world to the infinite universe, stand out clearly in the works of a few great thinkers who, in deep understanding of its primary importance, have given their full attention to the fundamental problem of the structure of the world. It is with them, and their works, that we shall be concerned here, all the more so as they present themselves to us in the form of a closely connected discussion.

I. The Sky and the Heavens

Nicholas of Cusa

& Marcellus Palingenius

The conception of the infinity of the universe, like everything else or nearly everything else, originates, of course, with the Greeks; and it is certain that the speculations of the Greek thinkers about the infinity of space and the multiplicity of worlds have played an important part in the history we shall be dealing with.[4] It seems to me, however, impossible to reduce the history of the infinitization of the universe to the rediscovery of the world-view of the Greek atomists which became better known through the newly discovered Lucretius [5] or the newly translated Diogenes Laertius.[6] We must not forget that the infinitist conceptions of the Greek atomists were rejected by the main trend, or trends, of Greek philosophical and scientific thought — the Epicurean tradition was not a scientific one [7] — and that for this very reason, though never forgotten, they could not be accepted by the mediaevals.

We must not forget, moreover, that " influence " is not a simple, but on the contrary, a very complex, bilateral relation. We are not influenced by everything we read or learn. In one sense, and perhaps the deepest, we ourselves determine the influences we are submitting to; our intellectual ancestors are by no means given to, but are freely chosen by, us. At least to a large extent.

How could we explain otherwise that, in spite of their great popularity, neither Diogenes nor even Lucretius had, for more than a century, any influence on the fifteenth century's cosmological thinking? The first man to take Lucretian cosmology seriously was Giordano Bruno. Nicholas of Cusa — it is true that it is not certain whether at the time when he wrote his *Learned Ignorance* (1440) he knew the *De rerum natura* — does not seem to have paid much attention to it. Yet it was Nicholas of Cusa, the last great philosopher of the dying Middle Ages, who first rejected the mediaeval cosmos-conception and to whom, as often as not, is ascribed the merit, or the crime, of having asserted the infinity of the universe.

It is indeed in such a way that he was interpreted by Giordano Bruno, by Kepler and, last but not least, by Descartes, who in a well-known letter to his friend Chanut (Chanut reports some reflections of Christina of Sweden, who doubted whether, in the indefinitely extended universe of Descartes, man could still occupy the central position that, according to the teaching of religion, was given to him by God in the creation of the world) tells the latter that after all " the Cardinal of Cusa and several other Divines have supposed the world to be infinite, without ever being reproached by the Church; on the contrary, it is believed that to make His works appear very great is to honor God."[8] The Cartesian interpretation of the teaching of Nicholas of Cusa is rather plausible as, indeed, Nicholas of Cusa denies the finitude of the world and its enclosure by the walls of the heavenly spheres. But he

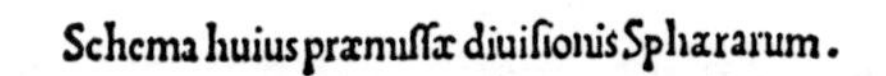

FIGURE 1

Typical pre-Copernican diagram of the universe

(from the 1539 edition of Peter Apian's *Cosmographia*)

does not assert its positive infinity; as a matter of fact he avoids as carefully and as consistently as Descartes himself the attribution to the universe of the qualification " infinite," which he reserves for God, and for God alone. His universe is not infinite (*infinitum*) but " interminate " (*interminatum*), which means not only that it is boundless and is not terminated by an outside shell, but also that it is not " terminated " in its constituents, that is, that it utterly lacks precision and strict determination. It never reaches the " limit "; it is, in the full sense of the word, *indetermined.* It cannot, therefore, be the object of total and precise knowledge, but only that of a partial and conjectural one.[9] It is the recognition of this necessarily partial — and relative — character of our knowledge, of the impossibility of building a univocal and objective representation of the universe, that constitutes — in one of its aspects — the *docta ignorantia,* the *learned ignorance,* advocated by Nicholas of Cusa as a means of transcending the limitations of our rational thought.

The world-conception of Nicholas of Cusa is not based upon a criticism of contemporary astronomical or cosmological theories, and does not lead, at least in his own thinking, to a revolution in science. Nicholas of Cusa, though it has often been so claimed, is not a forerunner of Nicholas Copernicus. And yet his conception is extremely interesting and, in some of its bold assertions — or negations — it goes far beyond anything that Copernicus ever dared to think of.[10]

The universe of Nicholas of Cusa is an expression or a development (*explicatio*), though, of course, necessarily imperfect and inadequate, of God — imperfect and inadequate because it displays in the realm of multiplicity and separation what in God is present in an indissoluble and intimate unity (*complicatio*), a unity which embraces not only the different, but even the opposite, qualities or determinations of being. In its turn, every singular thing in the universe represents it — the universe — and thus also God, in its own particular manner; each in a manner different from that of all others, by " contracting " (*contractio*) the wealth of the universe in accordance with its own unique individuality.

The metaphysical and epistemological conceptions of Nicholas of Cusa, his idea of the coincidence of the opposites in the absolute which transcends them, as well as the correlative concept of learned ignorance as the intellectual act that grasps this relationship which transcends discursive, rational thought, follow and develop the pattern of the mathematical paradoxes involved in the infinitization of certain relations valid for finite objects. Thus, for instance, nothing is more opposed in geometry than " straightness " and " curvilinearity "; and yet in the infinitely great circle the circumference coincides with the tangent, and in the infinitely small one, with the diameter. In both cases, moreover, the center loses its unique, determinate position; it coincides with the circumference; it is nowhere, or everywhere. But " great " and " small " are themselves a pair of opposed concepts that are valid and meaningful only in the realm of finite quantity, the realm of relative being, where there are no " great " or " small " objects, but only " greater " and " smaller " ones, and where, therefore, there is no " greatest," as well as no " smallest." Compared with the infinite there is nothing that is greater or smaller than anything else. The absolute, infinite maximum does not, any more than the absolute, infinite minimum, belong to the series of the great and small. They are outside it, and therefore, as Nicholas of Cusa boldly concludes, they coincide.

Another example can be provided by kinematics. No two things, indeed, are more opposed than motion and rest. A body in motion is never in the same place; a body at rest is never outside it. And yet a body moving with infinite velocity along a circular path will always be in the place of its departure, and at the same time will always be elsewhere, a good proof that motion is a relative concept embracing the oppositions of " speedy " and " slow." Thus it follows that, just as in the sphere of purely geometrical quantity, there is no minimum and no maximum of motion, no slowest and no quickest, and that the absolute maximum of velocity (infinite speed) as well as its absolute minimum (infinite slowness or rest) are both outside it, and, as we have seen, coincide.

Nicholas of Cusa is well aware of the originality of his thought and even more so of the rather paradoxical and strange character of the conclusion to which he is led by *learned ignorance.*[11]

> It is possible [he states] that those who will read things previously unheard of, and now established by *Learned Ignorance*, will be astonished.

Nicholas of Cusa cannot help it: it has, indeed, been established by *learned ignorance* [12]

> . . . that the universe is triune; and that there is nothing that is not a unity of potentiality, actuality and connecting motion; that no one of these can subsist absolutely without the other; and that all these are in all [things] in different degrees, so different that in the universe no two [things]

> can be completely equal to each other in everything. Accordingly, if we consider the diverse motions of the [celestial] orbs, [we find that] it is impossible for the machine of the world to have any fixed and motionless center; be it this sensible earth, or the air, or fire or anything else. For there can be found no absolute minimum in motion, that is, no fixed center, because the minimum must necessarily coincide with the maximum.

Thus the centrum of the world coincides with the circumference and, as we shall see, it is not a physical, but a metaphysical " centrum," which does not belong to the world. This " centrum," which is the same as the " circumference," that is, beginning and end, foundation and limit, the " place " that " contains " it, is nothing other than the Absolute Being or God.

Indeed, pursues Nicholas of Cusa, curiously reversing a famous Aristotelian argument in favour of the limitation of the world: [13]

> The world has no circumference, because if it had a center and a circumference, and thus had a beginning and end in itself, the world would be limited in respect to something else, and outside the world there would be something other, and space, things that are wholly lacking in truth. Since, therefore, it is impossible to enclose the world between a corporeal centrum and a circumference, it is [impossible for] our reason to have a full understanding of the world, as it implies the comprehension of God who is the center and the circumference of it.

Thus,[14]

> . . . though the world is not infinite, yet it cannot be conceived as finite, since it has no limits between which it is confined. The earth, therefore, which cannot be the center,

cannot be lacking in all motion; but it is necessary that it move in such a way that it could be moved infinitely less. Just as the earth is not the center of the world, so the sphere of the fixed stars is not its circumference, though if we compare the earth to the sky, the earth appears to be nearer to the center, and the sky to the circumference. The earth therefore is not the center, neither of the eighth nor of [any] other sphere, nor does the rising of the six signs [of the Zodiac] above the horizon imply that the earth is in the center of the eighth sphere. For even if it were somewhat distant from the center and outside the axis, which traverses the poles, so that in one part it would be elevated towards one pole, and in the other [part] depressed towards the other, nevertheless it is clear that, being at such a great distance from the poles and the horizon being just as vast, men would see only half of the sphere [and therefore believe themselves to be in its center].

Furthermore the very center of the world is no more inside the earth than outside it; for neither this earth, nor any other sphere, has a center; indeed, the center is a point equidistant from the circumference; but it is not possible that there be a true sphere or circumference such that a truer, and more precise one, could not be possible; a precise equidistance of divers [objects] cannot be found outside of God, for He alone is the infinite equality. Thus it is the blessed God who is the center of the world; He is the center of the earth and of all the spheres, and of all [the things] that are in the world, as He is at the same time the infinite circumference of all. Furthermore, there are in the sky no immovable, fixed poles, though the sky of the fixed stars appears by its motion to describe circles graduated in magnitude, lesser than the colures or than the equinoctials and also circles of an intermediate [magnitude]; yet, as a matter of fact, all the parts of the sky must move, though unequally

> in comparison with the circles described by the motion of the fixed stars. Thus, as certain stars appear to describe the maximal circle, so certain [others], the minimal, but there is no star that does not describe any. Therefore, as there is no fixed pole in the sphere, it is obvious that neither can there be found an exact mean, that is, a point equidistant from the poles. There is therefore no star in the eighth sphere which by [its] revolution would describe a maximal circle, because it would have to be equidistant from the poles which do not exist, and accordingly [the star] that would describe the minimal circle does not exist either. Thus the poles of the spheres coincide with the center and there is no other center than the pole, that is, the blessed God Himself.

The exact meaning of the conception developed by Nicholas of Cusa is not quite clear; the texts that I have quoted could be — and have been — interpreted in many different ways which I will not examine here. As for myself, I believe that we can understand it as expressing, and as stressing, the lack of precision and stability in the created world. Thus, there are no stars *exactly* on the poles, or on the equator of the celestial sphere. There is no *fixed constant* axis; the eighth, as well as all the other spheres, perform their revolutions around axes that continuously shift their positions. Moreover, these spheres are by no means *exact*, mathematical ("true") spheres, but only something which we should today call "spheroids"; accordingly, they have no center, in the precise meaning of this term. It follows therefore that neither the earth, nor anything else, can be placed in this center, which does not exist, and that thus nothing in this world can be completely and absolutely at rest.

I do not believe we can go further than that and attribute to Nicholas of Cusa a purely relativistic conception of space, such as, for instance, Giordano Bruno imputes to him. Such a conception implies the denial of the very existence of celestial orbs and spheres, which we cannot ascribe to Nicholas of Cusa.

Yet, in spite of this retention of the spheres, there is a good deal of relativism in Nicholas of Cusa's world-view. Thus he continues: [15]

> But we cannot discover motion unless it be by comparison with something fixed, that is [by referring it to] the poles or the centers and assuming them in our measurements of the motions [as being at rest]; it follows therefrom that we are always using conjectures, and err in the results [of our measurements]. And [if] we are surprised when we do not find the stars in the places where they should be according to the ancients, [it is] because we believe [wrongly] that they were right in their conceptions concerning the centers and poles as well as in their measurements.

It seems, then, that for Nicholas of Cusa the lack of agreement between the observations of the ancients and those of the moderns has to be explained by a change in the position of the axis (and poles), and, perhaps, by a shift in that of the stars themselves.

From all this, that is, from the fact that nothing in the world can be at rest, Nicholas of Cusa concludes:

> . . . it is obvious that the earth moves. And because from the motion of the comets, of the air and of fire, we know by experience that the elements move, and [that] the moon [moves] less from the Orient to the Occident than Mercury or Venus, or the sun, and so on, it follows that the earth

[considered as an element] moves less than all the others; yet [considered] as a star, it does not describe around the center or the pole a minimal circle, nor does the eighth sphere, or any other, describe the maximal, as has already been proved.

You have now to consider attentively what follows: just as the stars move around the conjectural poles of the eighth sphere, so also do the earth, the moon and the planets move in various ways and at [different] distances around a pole, which pole we have to conjecture as being [in the place] where you are accustomed to put the center. It follows therefrom that though the earth is, so to speak, the star which is nearer the central pole [than the others] it still moves, and yet does not describe in [its] motion the minimum circle, as has been shown *supra.* Moreover, neither the sun, nor the moon, nor any sphere—though to us it seems otherwise—can in [its] motion describe a true circle, because they do not move around a fixed base. Nowhere is there a true circle such that a truer one would not be possible, nor is [anything] ever at one time [exactly] as at another, neither does it move in a precisely equal [manner], nor does it describe an equally perfect circle, though we are not aware of it.

It is rather difficult to say precisely what kind of motion is ascribed to the earth by Nicholas of Cusa. In any case, it does not seem to be any of those that Copernicus was to attribute to it: it is neither the daily rotation around its axis, nor the annual revolution around the sun, but a kind of loose orbital gyration around a vaguely determined and constantly shifting center. This motion is of the same nature as that of all other celestial bodies, the sphere of the fixed stars included, though the slowest of them all, that of the sphere of the fixed stars being the quickest.

As for Nicholas of Cusa's assertions (quite unavoidable from his epistemological premises) that there is nowhere a precise circular orb or a precisely uniform motion, they must be interpreted as implying immediately (though he does not say it explicitly, it is clearly enough suggested by the context) that not only the factual content, but the very ideal of Greek and mediaeval astronomy, that is, the reduction of celestial motions to a system of interlocking uniform circular ones which would "save" the phenomena by revealing the permanent stability of the real behind the seeming irregularity of the apparent, is fallacious and must be abandoned.

Yet Nicholas of Cusa goes even further and, drawing the (penultimate) conclusion from the relativity of the perception of space (direction) and motion, he asserts that as the world-image of a given observer is determined by the place he occupies in the universe; and as none of these places can claim an absolutely privileged value (for instance, that of being the center of the universe), we have to admit the possible existence of different, equivalent world-images, the relative — in the full sense of the word — character of each of them, and the utter impossibility of forming an objectively valid representation of the universe.[16]

> Consequently, if you want to have a better understanding of the motion of the universe, you must put together the center and the poles, with the aid of your imagination as far as you can; for if somebody were on the earth, under the arctic pole, and somebody else on the arctic pole, then just as to the man on the earth the pole will appear to be

in the zenith, to the man on the pole it is the center that would appear to be in the zenith. And as the antipodes have, like ourselves, the sky above them, so to those who are in the poles (in both), the earth will appear to be in the zenith, and wherever the observer be he will believe himself to be in the center. Combine thus these diverse imaginations, making the center into the zenith and *vice versa,* and then, with the intellect, which alone can practise *learned ignorance,* you will see that the world and its motion cannot be represented by a figure, because it will appear almost as a wheel within a wheel, and a sphere within a sphere, having nowhere, as we have seen, either a center or a circumference.

The ancients [continues Nicholas of Cusa [17]] did not arrive at the things that we have brought forth, because they were deficient in *learned ignorance.* But for us it is clear that this earth really moves, though it does not appear to us to do so, because we do not apprehend motion except by a certain comparison with something fixed. Thus if a man in a boat, in the middle of a stream, did not know that the water was flowing and did not see the bank, how would he apprehend that the boat was moving? [18] Accordingly, as it will always seem to the observer, whether he be on the earth, or on the sun or on another star, that he is in the *quasi*-motionless center and that all the other [things] are in motion, he will certainly determine the poles [of this motion] in relation to himself; and these poles will be different for the observer on the sun and for the one on the earth, and still different for those on the moon and Mars, and so on for the rest. Thus, the fabric of the world (*machina mundi*) will *quasi* have its center everywhere and its circumference nowhere, because the circumference and the center are God, who is everywhere and nowhere.

It must be added that this earth is not spherical, as

> some have said, though it tends towards sphericity; indeed, the shape of the world is contrasted in its parts, as well as its motion; but when the infinite line is considered as contracted in such a way that, as contracted, it could not be more perfect or more spacious, then it is circular, and the corresponding corporeal figure [is the] spherical one. For all motion of the parts is towards the perfection of the whole; thus heavy bodies [move] towards the earth, and light ones [move] upward, earth towards earth, water towards water, fire towards fire; accordingly, the motion of the whole tends as far as it can towards the circular, and all shapes towards the spherical one, as we see in the parts of animals, in trees, and in the sky. But one motion is more circular and more perfect than another, and it is the same with shapes.

We cannot but admire the boldness and depth of Nicholas of Cusa's cosmological conceptions which culminate in the astonishing transference to the universe of the pseudo-Hermetic characterization of God: "a sphere of which the center is everywhere, and the circumference nowhere."[19] But we must recognize also that, without going far beyond him, it is impossible to link them with astronomical science or to base upon them a "reformation of astronomy." This is probably why his conceptions were so utterly disregarded by his contemporaries, and even by his successors for more than a hundred years. No one, not even Lefèvre d'Etaples who edited his works, seems to have paid much attention to them,[20] and it was only *after* Copernicus — who knew the works of Nicholas of Cusa, at least his treatise on the quadrature of the circle, but does not seem to have been influenced by him[21] — and even *after* Giordano Bruno, who drew his chief inspiration from him, that Nicholas of Cusa achieved fame as a forerunner of Copernicus, and even of Kepler, and could be quoted by Descartes as an advocate of the infinity of the world.

It is rather tempting to follow the example of these illustrious admirers of Nicholas of Cusa, and to read into him all kinds of anticipations of later discoveries, such, for instance, as the flattened form of the earth, the elliptic trajectories of the planets, the absolute relativity of space, the rotation of the heavenly bodies upon their axes.

Yet we must resist this temptation. As a matter of fact, Nicholas of Cusa does not assert anything of the kind. He *does* believe in the existence and also in the motion of heavenly spheres, that of the fixed stars being the quickest of all, as well as in the existence of a central region of the universe around which it moves as a whole, conferring this motion on all its parts. He *does not* assign a rotary motion to the planets, not even to this our earth. He *does not* assert the perfect uniformity of space. Moreover, in deep opposition to the fundamental inspiration of the founders of modern science and of the modern world-view, who, rightly or wrongly, tried to assert the panarchy of mathematics, he denies the very possibility of the mathematical treatment of nature.

We must now turn our attention to another aspect of the cosmology of Nicholas of Cusa, historically perhaps the most important: his rejection of the hierarchical structure of the universe, and, quite particularly, his denial — together with its central position — of the uniquely low and despicable position assigned to the earth by traditional cosmology. Alas, here too, his deep metaphysical intuition is marred by scientific conceptions that were not in advance of but rather behind his time, such as, for instance, the attribution to the moon, and even to the earth, of a light of their own.[22]

> The shape of the earth is noble and spherical, and its motion is circular, though it could be more perfect. And since in the world there is no maximum in perfections, motions and figures (as is evident from what has already been said) it is not true that this earth is the vilest and lowest [of the bodies of the world], for though it seems to be more central in relation to the world, it is also, for the same reason, nearer to the pole. Neither is this earth a proportional, or aliquot part of the world, for as the world has neither maximum, nor minimum, neither has it a moiety, nor aliquot parts, any more than a man or an animal [has them]; for the hand is not an aliquot part of the man, though its weight seems to have a proportion to the body, just as it does to the dimension and the figure. Nor is the dark colour [of the earth] an argument for its baseness, because to an observer on the sun, it [the sun] would not appear as brilliant as it does to us; indeed, the body of the sun must have a certain more central part, a *quasi* earth, and a certain circumferential *quasi*-fiery lucidity, and in the middle a *quasi*-watery cloud and clear air, just as this earth has its elements.[23] Thus someone outside the region of fire would see [the earth as] a brilliant star, just as to us, who are outside the region of the sun, the sun appears very luminous.

Having thus destroyed the very basis of the opposition of the " dark " earth and the " luminous " sun by establishing the similarity of their fundamental structure, Nicholas proclaims victoriously: [24]

> The earth is a noble star, which has a light and a heat and an influence of its own, different from those of all other

stars; every [star] indeed differs from every other in light, nature and influence; and thus every star communicates its light and influence to [every] other; not intentionally, for stars move and glitter only in order to exist in a more perfect manner: the participation arises as a consequence; just as light shines by its own nature, not in order that I may see it.

Indeed, in the infinitely rich and infinitely diversified and organically linked-together universe of Nicholas of Cusa, there is no center of perfection in respect to which the rest of the universe would play a subservient part; on the contrary, it is by being themselves, and asserting their own natures, that the various components of the universe contribute to the perfection of the whole. Thus the earth in its way is just as perfect as the sun, or the fixed stars. Accordingly, Nicholas of Cusa continues: [25]

> It must not be said either that, because the earth is smaller than the sun, and receives an influence from it, it is also more vile; for the whole region of the earth, which extends to the circumference of the fire, is large. And though the earth is smaller than the sun, as is known to us from its shadow and the eclipses, still we do not know whether the region of the sun is greater or smaller than the region of the earth; however, they cannot be precisely equal, as no star can be equal to another. Nor is the earth the the smallest star, for it is larger than the moon, as we are taught by the experience of the eclipses, and even, as some people say, larger than Mercury, and possibly than some other stars. Thus the argument from the dimension to the vileness is not conclusive.

Nor can it be argued that the earth is less perfect than the sun and the planets because it receives an influence from them: it is, as a matter of fact, quite possible that it influences them in its turn: [26]

> It is clear therefore that it is not possible for human knowledge to determine whether the region of the earth is in a degree of greater perfection or baseness in relation to the regions of the other stars, of the sun, the moon and the rest.

Some of the arguments in favour of the relative perfection of the earth are rather curious. Thus, being convinced that the world is not only unlimited but also everywhere populated, Nicholas of Cusa tells us that no conclusion as to the imperfection of the earth can be drawn from the alleged imperfection of its inhabitants, a conclusion that nobody, as far as I know, ever made, at least not in his time. Be that as it may, in any case Nicholas of Cusa asserts that,[27]

> . . . it cannot be said that this place of the world [is less perfect because it is] the dwelling-place of men, and animals, and vegetables that are less perfect than the inhabitants of the region of the sun and of the other stars. For although God is the center and the circumference of all the stellar regions, and although in every region inhabitants of diverse nobility of nature proceed from Him, in order that such vast regions of the skies and of the stars should not remain void, and that not only this earth be inhabited by lesser beings, still it does not seem that, according to the order of nature, there could be a more noble or more perfect nature than the intellectual nature which dwells here on this earth as in its region, even if there are in the other stars inhabitants belonging to another genus: man indeed does not desire another nature, but only the perfection of his own.

But, of course, we have to admit that in the same *genus* there may be several different *species* which embody the same common nature in a more, or less, perfect way. Thus it seems to Nicholas of Cusa rather reasonable to conjecture that the inhabitants of the sun and the moon are placed higher on the scale of perfection than ourselves: they are more intellectual, more spiritual than we, less material, less burdened by flesh.

And, finally, the great argument from change and corruptibility to baseness is declared by Nicholas of Cusa as having no more value than the rest. For [28] "since there is one universal world, and since all the particular stars influence each other in a certain proportion," there is no reason to suppose that change and decay occur only here, on the earth, and not everywhere in the universe. Nay, we have every reason to suppose — though of course we cannot *know* it — that it is everywhere the same, the more so as this corruption, which is presented to us as the particular feature of terrestrial being, is by no means a real destruction, that is, total and absolute loss of existence. It is, indeed, loss of that particular form of existence. But fundamentally it is not so much outright disappearance as dissolution, or resolution, of a being into its constitutive elements and their reunification into something else, a process that may take place — and probably does take place — in the whole universe just because the ontological structure of the world is, fundamentally, everywhere the same. Indeed it expresses everywhere in the same temporal, that is, mutable and changing, manner the immutable and eternal perfection of the Creator.

As we see, a new spirit, the spirit of the Renaissance breathes in the work of Cardinal Nicholas of Cusa. His world is no longer the medieval cosmos. But it is not yet, by any means, the infinite universe of the moderns.

The honor of having asserted the infinity of the universe has also been claimed by modern historians for a sixteenth century writer, Marcellus Stellatus Palingenius,[29] author of a widely read and very popular book, *Zodiacus vitae*, which was published in Venice, in Latin, in 1534 (and translated into English in 1560); but, in my opinion, with even less reason than in the case of Nicholas of Cusa.

Palingenius, who is deeply influenced by the Neoplatonic revival of the fifteenth century and who therefore rejects the absolute authority of Aristotle, though, at other times, he quotes him with approval, may have had some knowledge of Nicholas of Cusa's world-view and have been encouraged by his example in denying the finitude of creation. Yet it is not certain, since, with the exception of the rather energetic assertion of the impossibility of imposing a limit on God's creative action, we do not find in his teaching any reference to the particular tenets of the cosmology of Nicholas of Cusa.

Thus, for instance, when in discussing the general structure of the universe he tells us [30]

> But some have thought that every starre a worlde we may call,
> The earth they count a darkened starre, whereas the least of all.

it is obvious that it is not Nicholas of Cusa, but the ancient Greek cosmologists that he has in mind. It is to be noted, moreover, that Palingenius does not share their views. His own are quite different. He does not make the earth a star. On the contrary he maintains consistently the opposition between the terrestrial and the celestial regions; and it is just the imperfection of the former that leads him to the denial of its being the only populated place in the world.

Indeed,[31]

. . . we see
The Seas and earth with sundry sorts, of creatures full to bee.
Shall then the heavens cleare be thought, as void and empty made
O rather void and empty mindes, that thus yourselves persuade.

It is clear that we cannot share the errors of these "empty mindes." It is clear, too, that[32]

. . . creatures doth the skies containe, and every Starre beside
Be heavenly townes and seates of Saints, where Kings and Commons bide
Not shapes and shadows vain of things (as we have present here)
But perfect Kings and people eke, all things are perfect there.

Yet Palingenius does not assert the infinity of the world. It is true that, applying consistently the principle to which Professor Lovejoy has given the name of *principle of plenitude*,[33] he denies the finitude of God's creation, and says:[34]

A sorte there are that do suppose, the end of everything
Above the heavens to consist, and farther not to spring.
So that beyond them nothing is: and that above the *skies*

The *Nature* never powre to clime, but there amazed lies.
Which unto me appeareth false: and reason does me teach,
For if the ende of all be there, where skies no farther reach
Why hath not God created more? Because he had not skill
How more to make, his cunning staied and broken of his will?
Or for because he had not power? but truth both these denies,
For power of God hath never end, nor bounds his knowledge ties.
But in the *State Diuine* of *God* and *Glorious maiestie*
We must believe is nothing vaine since Godliest is the same:
This God what so ever he could doe assuredly did frame,
Least that his vertue were in vaine, and never should be hid.
But since he could make endlesse things, it never must be thought he did.

Nevertheless he maintains the finitude of the *material* world, enclosed and encompassed by the eight heavenly spheres: [35]

But *learned Aristotle* sayth there can no body bee,
But that it must of bondes consist: to this I do agree,
Because above the skies no kinde of body do we place,
But light most pure, of bodye voide, such light as doth deface
And farre excell our shining Sunne, such light as comprehend
Our eyes cannot, and endlesse light that God doth from him send.
Wherein together with their King the Sprites that are more hie
Doe dwell, the meaner sorte beneath the skies doe alwaies lie.
Therefore the reigne and position of the world consists in three,

Celestiall, Subcelestiall which with limits compast bee:
The Rest no boundes may comprehend which bright aboue the Skye
Doth shine with light most wonderfull. But here some will replye
That without body is no light, and so by this deny
That light can never there be found Above the Heavens by.

But Palingenius does not accept this theory which makes light dependent on matter and thus material itself. In any case, even if it were so for natural, physical light, it is certain that it is not the case for God's supernatural one. Above the starry heavens there are no *bodies*. But light and immaterial being can well be — and are — present in the supernatural, boundless supracelestial region.

Thus it is God's heaven, not God's world, that Palingenius asserts to be infinite.

II. The New Astronomy and the New Metaphysics

N. Copernicus
Th. Digges
G. Bruno
& W. Gilbert

Palingenius and Copernicus are practically contemporaries. Indeed, the *Zodiacus vitae* and the *De revolutionibus orbium cœlestium* must have been written at about the same time. Yet they have nothing, or nearly nothing, in common. They are as far away from each other as if they were separated by centuries.

As a matter of fact, they are, indeed, separated by centuries, by all those centuries during which Aristotelian cosmology and Ptolemaic astronomy dominated Western thought. Copernicus, of course, makes full use of the mathematical technics elaborated by Ptolemy — one of the greatest achievements of the human mind [1] — and yet, for his inspiration he goes back beyond him, and beyond Aristotle, to the golden age of Pythagoras and of Plato. He quotes Heraclides, Ecphantus and Hiketas, Philolaos and Aristarchus of Samos; and according to Rheticus, his pupil and mouthpiece, it is [2]

. . . following Plato and the Pythagoreans, the greatest

mathematicians of that divine age, that [he] thought that in order to determine the cause of the phenomena, circular motions have to be ascribed to the spherical earth.

I need not insist on the overwhelming scientific and philosophical importance of Copernican astronomy, which, by removing the earth from the center of the world and placing it among the planets, undermined the very foundations of the traditional cosmic world-order with its hierarchical structure and qualitative opposition of the celestial realm of immutable being to the terrestrial or sublunar region of change and decay. Compared to the deep criticism of its metaphysical basis by Nicholas of Cusa, the Copernican revolution may appear rather half-hearted and not very radical. It was, on the other hand, much more effective, at least in the long run; for, as we know, the immediate effect of the Copernican revolution was to spread skepticism and bewilderment [3] of which the famous verses of John Donne give such a striking, though somewhat belated, expression, telling us that the [4]

> . . . new Philosophy calls all in doubt,
> The Element of fire is quite put out;
> The Sun is lost, and th'earth, and no mans wit
> Can well direct him where to looke for it.
> And freely men confesse that this world's spent,
> When in the Planets, and the Firmament
> They seeke so many new; then see that this
> Is crumbled out againe to his Atomies.
> 'Tis all in peeces, all cohaerence gone;
> All just supply, and all Relation.

To tell the truth, the world of Copernicus is by no means devoid of hierarchical features. Thus, if he asserts that it is not the skies which move, but the earth, it is not only because it seems irrational to move a tremendously big body instead of a relatively small one, " that which contains and locates and not that which is contained and located," but also because " the condition of *being at rest* is considered as nobler and more divine than that of *change* and *inconsistency*; the latter therefore, is more suited to the earth than to the universe." [5] And it is on account of its supreme perfection and value — source of light and of life — that the place it occupies in the world is assigned to the sun: the central place which, following the Pythagorean tradition and thus reversing completely the Aristotelian and mediaeval scale, Copernicus believes to be the best and the most important one.[6]

Thus, though the Copernican world is no more hierarchically structured (at least not fully: it has, so to say, two poles of perfection, the sun and the sphere of the fixed stars, with the planets in between), it is still a well-ordered world. Moreover, it is still a finite one.

This finiteness of the Copernican world may appear illogical. Indeed, the only reason for assuming the existence of the sphere of the fixed stars being their common motion, the negation of that motion should lead immediately to the negation of the very existence of that sphere; moreover, since, in the Copernican world, the fixed stars must be exceedingly big [7] — the smallest being larger than the whole *Orbis magnus* — the sphere of the fixed stars must be rather thick; it seems only reasonable to extend its volume indefinitely " upwards."

It is rather natural to interpret Copernicus this way, that is, as an advocate of the infinity of the world, all the more so as he actually raises the question of the possibility of an indefinite spatial extension beyond the stellar sphere, though refusing to treat that problem as not scientific and turning it over to the philosophers. As a matter of fact, it is in this way that the Copernican doctrine was interpreted by Gianbattista Riccioli, by Huygens, and more recently by Mr. McColley.[8]

Though it seems reasonable and natural, I do not believe this interpretation to represent the actual views of Copernicus. Human thought, even that of the greatest geniuses, is never completely consequent and logical. We must not be astonished, therefore, that Copernicus, who believed in the existence of material planetary spheres because he needed them in order to explain the motion of the planets, believed also in that of a sphere of the fixed stars which he no longer needed. Moreover, though its existence did not explain anything, it still had some usefulness: the stellar sphere, which "embraced and contained everything and itself," held the world together and, besides, enabled Copernicus to assign a determined position to the sun.

In any case, Copernicus tells us quite clearly that[9]

> . . . the universe is spherical; partly because this form, being a complete whole, needing no joints, is the most perfect of all; partly because it constitutes the most spacious form which is thus best suited to contain and retain all things; or also because all discrete parts of the world, I mean the sun, the moon and the planets, appear as spheres.

True, he rejects the Aristotelian doctrine according to which "outside the world there is no body, nor place, nor empty space, in fact that nothing at all exists" because it seems to him " really strange that something could be enclosed by nothing " and believes that, if we admitted that " the heavens were infinite and bounded only by their inner concavity," then we should have better reason to assert " that there is nothing outside the heavens, because everything, whatever its size, is within them," [10] in which case, of course, the heavens would have to be motionless: the infinite, indeed, cannot be moved or traversed.

Yet he never tells us that the *visible world*, the world of the fixed stars, is infinite, but only that it is immeasurable (*immensum*), that it is so large that not only the earth compared to the skies is " as a point " (this, by the way, had already been asserted by Ptolemy), but also the whole orb of the earth's annual circuit around the sun; and that we do not and cannot know the limit, the dimension of the world. Moreover, when dealing with the famous objection of Ptolemy according to which " the earth and all earthly things if set in rotation would be dissolved by the action of nature," that is, by the centrifugal forces produced by the very great speed of its revolution, Copernicus replies that this disruptive effect would be so much stronger upon the heavens as their motion is more rapid than that of the earth, and that, " if this argument were true, the extent of the heavens would become infinite." In which case, of course, they would have to stand still, which, though finite, they do.

Thus we have to admit that, even if outside the world there were not nothing but space and even matter, nevertheless the *world* of Copernicus would remain a finite one, encompassed by a material sphere or orb, the sphere of the fixed stars — a sphere that has a centrum, a centrum occupied by the sun. It seems to me that there is no other way of interpreting the teaching of Copernicus. Does he not tell us that [11]

> . . . the first and the supreme of all [spheres] is the sphere of the fixed stars which contains everything and itself and which, therefore, is at rest. Indeed, it is the place of the world to which are referred the motion and the position of all other stars. Some [astronomers] indeed, have thought that, in a certain manner, this sphere is also subjected to change: but in our deduction of the terrestrial motion we have determined another cause why it appears so. [After the sphere of the fixed stars] comes Saturn, which performs its circuit in thirty years. After him, Jupiter, which moves in a duodecennial revolution. Then Mars which circumgirates in two years. The fourth place in this order is occupied by the annual revolution, which, as we have said, contains the Earth with the orb of the Moon as an epicycle. In the fifth place Venus revolves in nine months. Finally, the sixth place is held by Mercury, which goes around in the space of eighty days.
>
> But in the center of all resides the Sun. Who, indeed, in this most magnificent temple would put the light in another, or in a better place than that one wherefrom it could at the same time illuminate the whole of it? Therefore it is not improperly that some people call it the lamp of the world, others its mind, others its ruler. Trismegistus [calls it] the visible God, Sophocles' Electra, the All-Seeing. Thus, assuredly, as residing in the royal see the Sun governs the surrounding family of the stars.

We have to admit the evidence: the world of Copernicus is finite. Moreover, it seems to be psychologically quite normal that the man who took the first step, that of arresting the motion of the sphere of the fixed stars, hesitated before taking the second, that of dissolving it in boundless space; it was enough for one man to move the earth and to enlarge the world so as to make it immeasurable — *immensum*; to ask him to make it infinite is obviously asking too much.

Great importance has been attributed to the enlargement of the Copernican world as compared to the mediaeval one — its diameter is at least 2000 times greater. Yet, we must not forget, as Professor Lovejoy has already pointed out,[12] that even the Aristotelian or Ptolemaic world was by no means that snug little thing that we see represented on the miniatures adorning the manuscripts of the Middle Ages and of which Sir Walter Raleigh gave us such an enchanting description.[13] Though rather small by our astronomical standards, and even by those of Copernicus, it was in itself sufficiently big not to be felt as built to man's measure: about 20,000 terrestrial radii, such was the accepted figure, that is, about 125,000,000 miles.

Let us not forget, moreover, that, by comparison with the infinite, the world of Copernicus is by no means greater than that of mediaeval astronomy; they are both as nothing, because *inter finitum et infinitum non est proportio*. We do not approach the infinite universe by increasing the dimensions of our world. We may make it as large as we want: that does *not* bring us any nearer to it.[14]

Notwithstanding this, it remains clear that it is somewhat easier, psychologically if not logically, to pass from a very large, immeasurable and ever-growing world to an infinite one than to make this jump starting with a rather big, but still determinably limited sphere: the world-bubble has to swell before bursting. It is also clear that by his reform, or revolution, of astronomy Copernicus removed one of the most valid scientific objections against the infinity of the universe, based, precisely, upon the empirical, common-sense fact of the motion of the celestial spheres.

The infinite cannot be traversed, argued Aristotle; now the stars turn around, therefore . . . But the stars do not turn around; they stand still, therefore . . . It is thus not surprising that in a rather short time after Copernicus some bold minds made the step that Copernicus refused to make, and asserted that the celestial sphere, that is the sphere of the fixed stars of Copernican astronomy, does not exist, and that the starry heavens, in which the stars are placed at different distances from the earth, "extendeth itself infinitely up."

It has been commonly assumed until recent times that it was Giordano Bruno who, drawing on Lucretius and creatively misunderstanding both him and Nicholas of Cusa,[15] first made this decisive step. Today, after the discovery by Professor Johnson and Dr. Larkey[16] — in 1934 — of the *Perfit Description of the Caelestiall Orbes according to the most aunciene doctrine of the Pythagoreans lately revived by Copernicus and by Geometricall Demonstrations approued*, which Thomas Digges, in 1576, added to the *Prognostication euerlasting* of his father Leonard Digges, this honor, at least partially, must be ascribed to him. Indeed, though different interpretations may be given of the text of Thomas Digges — and my own differs somewhat from that of Professor Johnson and Dr. Larkey — it is certain, in any case, that Thomas

Digges was the first Copernican to replace his master's conception, that of a closed world, by that of an open one, and that in his *Description,* where he gives a fairly good, though rather free, translation of the cosmological part of the *De revolutionibus orbium cœlestium,* he makes some rather striking additions. First, in his description of the orb of Saturn he inserts the clause that this orb is "of all others next vnto that infinite Orbe immouable, garnished with lights innumerable"; then he substitutes for the well-known Copernican diagram of the world another one, in which the stars are placed on the whole page, above as well as below the line by which Copernicus represented the *ultima sphaera mundi.* The text that Thomas Digges adds to his diagram is very curious. In my opinion, it expresses the hesitation and the uncertainty of a mind — a very bold mind — which on the one hand not only accepted the Copernican world-view, but even went beyond it, and which, on the other hand, was still dominated by the religious conception — or image — of a heaven located in space. Thomas Digges begins by telling us that:

> The orbe of the starres fixed infinitely up extendeth hit self in altitude sphericallye, and therefore immouable.

Yet he adds that this orbe is

> the pallace of felicitye garnished with perpetuall shininge glorious lightes innumerable, farr excelling our sonne both in quantity and qualitye.

And that it is

> the Court of the great God, the habitacle of the elect, and of the coelestiall angelles.

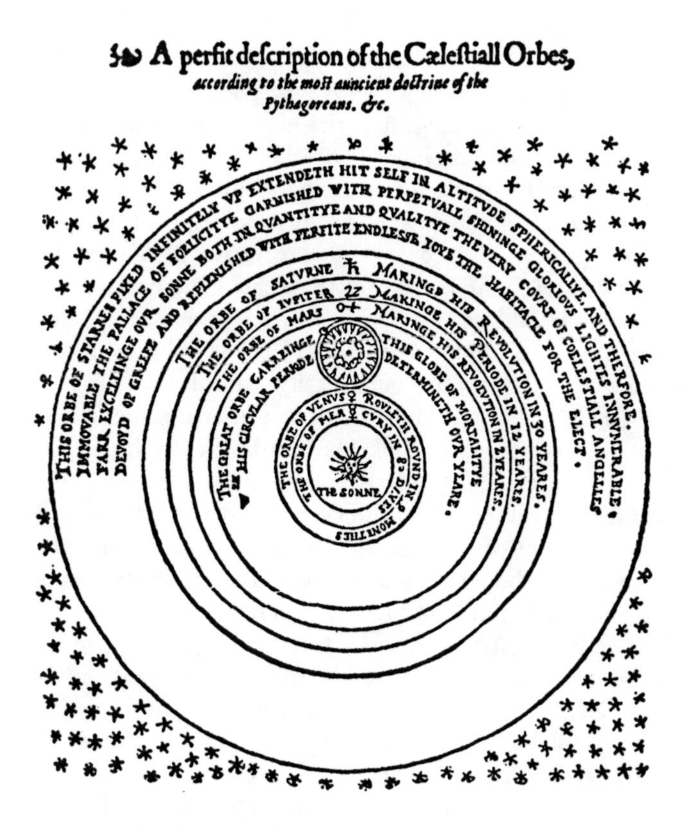

FIGURE 2

Thomas Digges's diagram of the infinite Copernican universe

(from *A Perfit Description of the Caelestiall Orbes*, 1576)

The text accompanying the diagram develops this idea: [18]

> Heerein can wee never sufficiently admire thys wonderfull and incomprehensible huge frame of goddes woorke proponed to our senses, seinge first the baull of ye earth wherein we moue, to the common sorte seemeth greate, and yet in respecte of the Moones Orbe is very small, but compared with the *Orbis magnus* wherein it is carried, it scarcely retayneth any sensible proportion, so merueillously is that Orbe of Annuall motion greater than this little darke starre wherein we liue. But that *Orbis magnus* beinge as is before declared but as a poynct in respect of the immensity of that immoueable heaven, we may easily consider what little portion of gods frame, our Elementare corruptible worlde is, but neuer sufficiently be able to admire the immensity of the Rest. Especially of that fixed Orbe garnished with lightes innumerable and reachinge vp in *Sphaericall altitude* without ende. Of which lightes Celestiall it is to bee thoughte that we only behoulde sutch as are in the inferioure partes of the same Orbe, and as they are hygher, so seeme they of lesse and lesser quantity, even tyll our syghte beinge not able farder to reache or conceyve, the greatest part rest by reason of their wonderfull distance inuisible vnto vs. And this may well be thought of vs to be the gloriouse court of ye great god, whose vnsercheable works inuisible we may partly by these his visible coniecture, to whose infinit power and maiesty such an infinit place surmountinge all other both in quantity and quality only is conueniente. But because the world hath so longe a time bin carried with an opinion of the earths stabilitye, as the contrary cannot but be nowe very imperswasible.

Thus, as we see, Thomas Digges puts his stars into a theological heaven; not into an astronomical sky. As a matter of fact, we are not very far from the conception of Palingenius — whom Digges knows and quotes — and, perhaps, nearer to him than to Copernicus. Palingenius, it is true, places his heaven above the stars, whereas Thomas Digges puts them into it. Yet he maintains the separation between our world — the world of the sun and the planets — and the heavenly sphere, the dwelling-place of God, the celestial angels, and the saints. Needless to say, there is no place for Paradise in the astronomical world of Copernicus.

That is the reason why, in spite of the very able defence of the priority rights of Digges made by Professor Johnson in his excellent book, *Astronomical Thought in Renaissance England,* I still believe that it was Bruno who, for the first time, presented to us the sketch, or the outline, of the cosmology that became dominant in the last two centuries, and I cannot but agree with Professor Lovejoy, who in his classical *Great Chain of Being* tells us that,[19]

> Though the elements of the new cosmography had, then, found earlier expression in several quarters, it is Giordano Bruno who must be regarded as the principal representative of the doctrine of the decentralised, infinite and infinitely populous universe; for he not only preached it throughout western Europe with the fervour of an evangelist, but also first gave a thorough statement of the grounds on which it was to gain acceptance from the general public.

Indeed, never before has the essential infinitude of space been asserted in such an outright, definite and conscious manner.

Thus, already in the *La Cena de le Ceneri,*[20] where, by the way, Bruno gives the best discussion, and refutation, of the classical — Aristotelian and Ptolemaic — objections against the motion of the earth that were ever written before Galileo,[21] he proclaims that [22] " the world is infinite and that, therefore, there is no body in it to which it would pertain *simpliciter* to be in the center, or on the center, or on the periphery, or between these two extremes " of the world (which, moreover, do not exist), but only to be among other bodies. As for the world which has its cause and its origin in an infinite cause and an infinite principle, it must be infinitely infinite according to its corporeal necessity and its mode of being. And Bruno adds: [23]

> It is certain that . . . it will never be possible to find an even half-probable reason, why there should be a limit to this corporeal universe, and, consequently, why the stars, which are contained in its space, should be finite in number.

But we find the clearest, and most forceful, presentation of the new gospel of the unity and the infinity of the world in his vernacular dialogues *De l'infinito universo e mondi* and in his Latin poem *De immenso et innumerabilibus*.[24]

> There is a single general space, a single vast immensity which we may freely call Void: in it are innumerable globes like this on which we live and grow; this space we declare to be infinite, since neither reason, convenience, sense-perception nor nature assign to it a limit. For there is no reason, nor defect of nature's gifts, either of active or passive power, to hinder the existence of other worlds throughout space, which is identical in natural character with our own space, that is everywhere filled with matter or at least ether.[25]

We have, of course, heard nearly similar things from

Nicholas of Cusa. And yet we cannot but recognize the difference of accent. Where Nicholas of Cusa simply states the impossibility of assigning limits to the world, Giordano Bruno asserts, and rejoices in, its infinity: the superior determination and clarity of the pupil as compared to his master is striking.[26]

> To a body of infinite size there can be ascribed neither center nor boundary. For he who speaketh of emptiness, the void or the infinite ether, ascribeth to it neither weight nor lightness, nor motion, nor upper, nor lower, nor intermediate regions; assuming moreover that there are in this space those countless bodies such as our earth and other earths, our sun and other suns, which all revolve within this infinite space, through finite and determined spaces or around their own centres. Thus we on the earth say that the earth is in the centre; and all the philosophers ancient and modern of whatever sect will proclaim without prejudice to their own principles that here is indeed the centre.

Yet,

> Just as we say that we are at the centre of that [universally] equidistant circle, which is the great horizon and the limit of our own encircling ethereal region, so doubtlessly the inhabitants of the moon believe themselves at the centre [of a great horizon] that embraces the earth, the sun and the other stars, and is the boundary of the radii of their own horizon. Thus the earth no more than any other world is at the centre; moreover, no points constitute determined celestial poles for our earth, just as she herself is not a definite and determined pole to any other point of the ether, or of the world-space; and the same is true of all other bodies. From various points of view these may all be regarded either as centres, or as points on the circumference, as poles, or zeniths and so forth.

> Thus the earth is not in the centre of the Universe; it is central only to our surrounding space.

Professor Lovejoy, in his treatment of Bruno, insists on the importance for the latter of the principle of plenitude, which governs his thought and dominates his metaphysics.[27] Professor Lovejoy is perfectly right, of course: Bruno uses the principle of plenitude in an utterly ruthless manner, rejecting all the restrictions by which mediaeval thinkers tried to limit its applicability and boldly drawing from it all the consequences that it implies. Thus to the old and famous *questio disputata*: why did not God create an infinite world? — a question to which the mediaeval scholastics gave so good an answer, namely, denying the very possibility of an infinite creature — Bruno simply replies, and he is the first to do it: God did. And even: God could not do otherwise.

Indeed, Bruno's God, the somewhat misunderstood *infinitas complicata* of Nicholas of Cusa, could not but explicate and express himself in an infinite, infinitely rich, and infinitely extended world.[28]

> Thus is the excellence of God magnified and the greatness of his kingdom made manifest; he is glorified not in one, but in countless suns; not in a single earth, but in a thousand, I say, in an infinity of worlds.
>
> Thus not in vain the power of the intellect which ever seeketh, yea, and achieveth the addition of space to space, mass to mass, unity to unity, number to number, by the science that dischargeth us from the fetters of a most narrow kingdom and promoteth us to the freedom of a truly august realm, which freeth us from an imagined poverty and straineth to the possession of the myriad riches of so vast a space, of so worthy a field of so many cultivated

worlds. This science does not permit that the arch of the horizon that our deluded vision imagineth over the Earth and that by our phantasy is feigned in the spacious ether, shall imprison our spirit under the custody of a Pluto or at the mercy of a Jove. We are spared the thought of so wealthy an owner and subsequently of so miserly, sordid and avaricious a donor.

It has often been pointed out — and rightly, of course — that the destruction of the cosmos, the loss, by the earth, of its central and thus unique (though by no means privileged) situation, led inevitably to the loss, by man, of his unique and privileged position in the theo-cosmic drama of the creation, of which man was, until then, both the central figure and the stake. At the end of the development we find the mute and terrifying world of Pascal's " libertin,"[29] the senseless world of modern scientific philosophy. At the end we find nihilism and despair.

Yet this was not so in the beginning. The displacement of the earth from the centrum of the world was not felt to be a demotion. Quite the contrary: it is with satisfaction that Nicholas of Cusa asserts its promotion to the rank of the noble stars; and, as for Giordano Bruno, it is with a burning enthusiasm — that of a prisoner who sees the walls of his jail crumble — that he announces the bursting of the spheres that separated us from the wide open spaces and inexhaustible treasures of the ever-changing, eternal and infinite universe. Ever-changing! We are, once more, reminded of Nicholas of Cusa, and, once more, we have to state the difference of their fundamental world views — or world feelings. Nicholas of Cusa *states* that immutability can nowhere be found in the whole universe; Giordano Bruno goes far beyond this mere statement; for him motion and change are signs of perfection and not of a lack of it. An immutable universe would be a dead universe; a living one must be able to move and to change.[30]

> There are no ends, boundaries, limits or walls which can defraud or deprive us of the infinite multitude of things. Therefore the earth and the ocean thereof are fecund; therefore the sun's blaze is everlasting, so that eternally fuel is provided for the voracious fires, and moisture replenishes the attenuated seas. For from infinity is born an ever fresh abundance of matter.
>
> Thus Democritus and Epicurus, who maintained that everything throughout infinity suffereth renewal and restoration, understood these matters more truly than those who at all costs maintain a belief in the immutability of the Universe, alleging a constant and unchanging number of particles of identical material that perpetually undergo transformation, one into another.

The importance for Bruno's thought of the principle of plenitude cannot be overvalued. Yet there are in it two other features that seem to me to be of as great an importance as this principle. They are: (a) the use of a principle that a century later Leibniz — who certainly knew Bruno and was influenced by him — was to call *the principle of sufficient reason*, which supplements the principle of plenitude and, in due time, superseded it; and (b) the decisive shift (adumbrated indeed by Nicholas of Cusa) from sensual to intellectual cognition in its relation to thought (intellect). Thus, at the very beginning of his Dialogue on the *Infinite Universe and the Worlds*, Bruno (Philotheo) asserts that sense-perception, as such, is confused and erroneous and cannot be made the basis of scientific and philosophical knowledge. Later on he explains that whereas for sense-perception and imagination infinity is inaccessible and unrepresentable, for the intellect, on the contrary, it is its primary and most certain concept.[31]

> PHILOTHEO — No corporeal sense can perceive the infinite. None of our senses can be expected to furnish this conclusion; for the infinite cannot be the object of sense-perception; therefore he who demandeth to obtain this knowledge through the senses is like unto one who would desire to see with his eyes both substance and essence. And he who would deny the existence of a thing merely because it cannot be apprehended by the senses, nor is visible, would presently be led to the denial of his own substance and being. Wherefore there must be some measure in the demand for evidence from our sense-perception, for this we can accept only in regard to sensible objects, and even there it is not above all suspicion unless it cometh before the court aided by good judgment. It is the part of the intellect to judge yielding due weight to factors absent and separated by distance of time and by space intervals. And in this matter our sense-perception doth suffice us and doth yield us adequate testimony, since it is unable to gainsay us; moreover it advertiseth and confesseth its own feebleness and inadequacy by the impression it giveth us of a finite horizon, an impression moreover which is ever changing. Since then we have experience that sense-perception deceiveth us concerning the surface of this globe on which we live, much more should we hold suspect the impression it giveth us of a limit to the starry sphere.
>
> ELPINO — Of what use are the senses to us? tell me that.
>
> PHIL. — Solely to stimulate our reason, to accuse, to indi-

cate, to testify in part . . . truth is in but a very small degree derived from the senses as from a frail origin, and doth by no means reside in the senses.

ELP. — Where then?

PHIL. — In the sensible object as in a mirror; in reason, by process of argument and discussion. In the intellect, either through origin or by conclusion. In the mind, in its proper and vital form.

As for the principle of sufficient reason, Bruno applies it in his discussion of space and of the spatially extended universe. Bruno's space, the space of an infinite universe and at the same time the (somewhat misunderstood) infinite "void" of Lucretius, is perfectly homogeneous and similar to itself everywhere: indeed, how could the "void" space be anything but uniform — or *vice versa*, how could the uniform "void" be anything but unlimited and infinite? Accordingly, from Bruno's point of view, the Aristotelian conception of a closed innerworldly space is not only false, it is absurd.[32]

PHILOTHEO — If the world is finite and if nothing is beyond, I ask you *where* is the world? *Where* is the universe? Aristotle replieth: it is in itself. The convex surface of the primal heaven is universal space, which being the primal container is by nought contained.

FRACASTORO — The world then will be nowhere. Everything will be in nothing.

PHIL. — If thou wilt excuse thyself by asserting that where nought is, and nothing existeth, there can be no question of position in space, nor of beyond, nor outside, yet I shall in no wise be satisfied. For these are mere words and excuses which cannot form part of our thought. For it is wholly impossible that in any sense or fantasy (even

> though there may be various senses and various fantasies), it is, I say, impossible that I can with any true meaning assert that there existeth such a surface, boundary or limit, beyond which is neither body, nor empty space, even though God be there.

We can pretend, as Aristotle does, that this world encloses all being, and that outside this world there is nothing; *nec plenum nec vacuum.* But nobody can think, or even imagine it. " Outside " the world will be space. And this space, just as ours, will not be " void "; it will be filled with " ether."

Bruno's criticism of Aristotle (like that of Nicholas of Cusa) is, of course, wrong. He does not understand him and substitutes a geometrical " space " for the place-continuum of the Greek philosopher. Thus he repeats the classical objection: what would happen if somebody stretched his hand through the surface of the heaven? [33] And though he gives to this question a nearly correct answer (from the point of view of Aristotle), [34]

> BURCHIO — Certainly I think that one must reply to this fellow that if a person would stretch out his hand beyond the convex sphere of heaven, the hand would occupy no position in space, nor any place, and in consequence would not exist.

he rejects it on the perfectly fallacious ground that this " inner surface," being a purely mathematical conception, cannot oppose a resistance to the motion of a real body. Furthermore, even if it did, the problem of what is beyond it would remain unanswered: [35]

> PHILOTHEO — Thus, let the surface be what it will, I must always put the question: what is beyond? If the reply

is: nothing, then I call that the void, or empty-ness. And such a Void or Emptiness hath no measure nor outer limit, though it hath an inner; and this is harder to imagine than is an infinite or immense universe. For if we insist on a finite universe, we cannot escape the void. And let us now see whether there can be such a space, in which is nought. In this infinite space is placed our universe (whether by chance, by necessity or by providence I do not now consider). I ask now whether this space which indeed containeth the world is better fitted to do so than is another space beyond?

FRACASTORO — It certainly appeareth to me not so. For where there is nothing there can be no differentiation; where there is no differentiation there is no distruction of quality and perhaps there is even less of quality where there is nought whatsoever.

Thus the space occupied by our world, and the space outside it, will be the same. And if they are the same, it is impossible that "outside" space should be treated by God in any different way from that which is "inside." We are therefore bound to admit that not only space, but also being in space is everywhere constituted in the same way, and that if in our part of the infinite space there is a world, a sun-star surrounded by planets, it is the same everywhere in the universe. Our world is not the universe, but only this *machina*, surrounded by an infinite number of other similar or analogous "worlds" — the worlds of star-suns scattered in the etheric ocean of the sky.[36]

Indeed, if it was, and is, possible for God to create a world in this our space, it is, and it was, just as possible for Him to create it elsewhere. But the uniformity of space — pure receptacle of being — deprives God of any reason to create it here, and not elsewhere. Accordingly, the limitation of God's creative action is unthinkable. In this case, the possibility implies actuality. The infinite world can be; therefore it must be; therefore it is.[37]

> For just as it would be ill that this our space were not filled, that is our world were not to exist, then, since the spaces are indistinguishable, it would be no less ill if the whole of space were not filled. Thus we see that the universe is of indefinite size and the worlds therein without number.

Or, as the Aristotelian adversary of Bruno, Elpino, now converted to his views, formulates it: [38]

> I declare that which I cannot deny, namely, that within infinite space either there may be an infinity of worlds similar to our own; or that this universe may have extended its capacity in order to contain many bodies such as those we name stars; or again that whether these worlds be similar or dissimilar to one another, it may with no less reason be well that one, than that another should exist. For the existence of one is no less reasonable than that of another; and the existence of many no less so than of one or of the other; and the existence of an infinity of them no less so than the existence of a large number. Wherefore, even as the abolition and non-existence of this world would be an evil, so would it be of innumerable others.

More concretely: [39]

> ELP. — There are then innumerable suns, and an infinite number of earths revolve around these suns, just as the seven we can observe revolve around this sun which is close to us.

PHIL. — So it is.

ELP. — Why then do we not see the other bright bodies which are the earths circling around the bright bodies which are suns? For beyond these we can detect no motion whatsoever; and why do all the other mundane bodies appear always (except those known as comets) in the same order and at the same distance?

Elpino's question is rather good. And the answer given to it by Bruno is rather good, too, in spite of an optical error of believing that, in order to be seen, the planets must be formed on the pattern of spherical mirrors and possess a polished, smooth, " watery " surface, for which, moreover, he is not responsible as it was common belief until Galileo: [40]

PHIL. — The reason is that we discern only the largest suns, immense bodies. But we do not discern the earths because, being much smaller they are invisible to us. Similarly, it is not impossible that other earths revolve around our sun and are invisible to us either on account of greater distance or smaller size, or because they have but little watery surface, or because such watery surface is not turned toward us and opposed to the sun, whereby it would be made visible as a crystal mirror which receiveth luminous rays; whence we perceive that it is not marvellous or contrary to nature that often we hear that the sun has been partially eclipsed though the moon hath not been interpolated between him and our sight. There may be innumerable watery luminous bodies — that is earths consisting in part of water circulating around the sun, besides those visible to us; but the difference in their orbits is indiscernible by us on account of their great distance, wherefore we perceive no difference in the very slow motion discernible of those visible above or beyond Saturn; still less doth there appear any order in the motion of all around the centre, whether we place our earth or our sun as that centre.

The question then arises whether the fixed stars of the heavens are really suns, and centers of worlds comparable to ours.[41]

ELP.—Therefore you consider that if the stars beyond Saturn are really motionless as they appear, then they are those innumerable suns or fires more or less visible to us around which travel their own neighbouring earths which are not discernible by us.

One would expect a positive answer. But for once Bruno is prudent: [42]

PHIL.—Not so for I do not know whether all or whether the majority is without motion, or whether some circle around others, since none hath observed them. Moreover they are not easy to observe, for it is not easy to detect the motion and progress of a remote object, since at a great distance change of position cannot easily be detected, as happeneth when we would observe ships in a high sea. But however that may be, the universe being infinite, there must be ultimately other suns. For it is impossible that heat and light from one single body should be diffused throughout immensity, as was supposed by Epicurus if we may credit what others relate of him. Therefore it followeth that there must be innumerable suns, of which many appear to us as small bodies; but that star will appear smaller which is in fact much larger than that which appeareth much greater.

The infinity of the universe thus seems to be perfectly

assured. But what about the old objection that the concept of infinity can be applied only to God, that is, to a purely spiritual, incorporeal Being, an objection which led Nicholas of Cusa — and later Descartes — to avoid calling their worlds " infinite," but only " interminate," or " indefinite "? Bruno replies that he does not deny, of course, the utter difference of the intensive and perfectly simple infinity of God from the extensive and multiple infinity of the world. Compared to God, the world is as a mere point, as a nothing.[43]

> PHIL. — We are then at one concerning the incorporeal infinite; but what preventeth the similar acceptability of the good, corporeal and infinite being? And why should not that infinite which is implicit in the utterly simple and indivisible Prime Origin rather become explicit in his own infinite and boundless image able to contain innumerable worlds, than become explicit within such narrow bounds? So that it appeareth indeed shameful to refuse to credit that this world which seemeth to us so vast may not in the divine regard appear a mere point, even a nullity?

Yet it is just that " nullity " of the world and of all the bodies that constitute it that implies its infinity. There is no reason for God to create one particular kind of beings in preference to another. The principle of sufficient reason reinforces the principle of plenitude. God's creation, in order to be perfect and worthy of the Creator, must therefore contain all that is possible, that is, innumerable individual beings, innumerable earths, innumerable stars and suns — thus we could say that God needs an infinite space in order to place in it this infinite world.

To sum up: [44]

PHIL.—This indeed is what I had to add; for, having pronounced that the Universe must itself be infinite because of the capacity and aptness of infinite space; on account also of the possibility and convenience of accepting the existence of innumerable worlds like to our own; it remaineth still to prove it. Now both from the circumstances of this efficient cause which must have produced the Universe such as it is, or rather, must ever produce it such as it is, and also from the conditions of our mode of understanding, we may easily argue that infinite space is similar to this which we see, rather than argue that it is that which we do not see either by example or by similitude, or by proportion, or indeed, by any effort of imagination which doth not finally destroy itself. Now to begin. Why should we, or could we imagine that divine power were otiose? Divine goodness can indeed be communicated to infinite things and can be infinitely diffused; why then should we wish to assert that it would choose to be scarce and to reduce itself to nought—for every finite thing is as nought in relation to the infinite? Why do you desire that centre of divinity which can (if one may so express it) extend indefinitely to an infinite sphere, why do you desire that it should remain grudgingly sterile rather than extend itself as a father, fecund, ornate and beautiful? Why should you prefer that it should be less, or indeed by no means communicated, rather than that it should fulfil the scheme of its glorious power and being? Why should infinite amplitude be frustrated, the possibility of an infinity of worlds be defrauded? Why should be prejudiced the excellency of the divine image which ought rather to glow in an unrestricted mirror, infinite, immense, according to the law of its being? . . . Why wouldst thou that God should in power, in act and in effect (which in him are identical) be determined as the limit of the convexity of a sphere rather than that he should be, as we may say, the undetermined limit of the boundless?

Let us not, adds Bruno, be embarrassed by the old objection that the infinite is neither accessible, nor understandable. It is the opposite that is true: the infinite is necessary, and is even the first thing that naturally *cadit sub intellectus.*

Giordano Bruno, I regret to say, is not a very good philosopher. The blending together of Lucretius and Nicholas of Cusa does not produce a very consistent mixture; and though, as I have already said, his treatment of the traditional objections against the motion of the earth is rather good, the best given to them before Galileo, he is a very poor scientist, he does not understand mathematics, and his conception of the celestial motions is rather strange. My sketch of his cosmology is, indeed, somewhat unilateral and not quite complete. As a matter of fact, Bruno's world-view is vitalistic, magical; his planets are animated beings that move freely through space of their own accord like those of Plato or of Pattrizzi. Bruno's is not a modern mind by any means. Yet his conception is so powerful and so prophetic, so reasonable and so poetic that we cannot but admire it and him. And it has — at least in its formal features — so deeply influenced modern science and modern philosophy, that we cannot but assign to Bruno a very important place in the history of the human mind.

I do not know whether Bruno had a great influence on his immediate contemporaries, or even whether he influenced them at all. Personally, I doubt it very much. He was, in his teaching, far ahead of his time.[45] Thus his influence seems to me to have been a delayed one. It was only *after* the great telescopic discoveries of Galileo that it was accepted and became a factor, and an important one, of the seventeenth century world-view.

Kepler, as a matter of fact, links Bruno with Gilbert and seems to suggest that it was from the former that the great British scientist received his belief in the infinity of the universe.

This is, of course, quite possible: the thorough criticism of the Aristotelian cosmology may have impressed Gilbert. Yet it would be the only point where the teaching of the Italian philosopher was accepted by him. There is, indeed, not much similarity (besides the animism, common to both) between the " magnetic philosophy " of William Gilbert and the metaphysics of Giordano Bruno. Professor Johnson believes that Gilbert was influenced by Digges, and that, having asserted the indefinite extension of the world " of which the limit is not known, and cannot be known," Gilbert, " to enforce his point, adopted without qualification Digges' idea that the stars were infinite in number, and located at varying and infinite distances from the center of the Universe." [46]

This is quite possible, too. Yet, if he adopted *this* idea of Digges, he completely rejected his predecessor's immersion of the celestial bodies into the theological heavens: he has nothing to tell us about the angels and the saints.

On the other hand, neither Bruno nor Digges succeeded in persuading Gilbert to accept, in its entirety, the astronomical theory of Copernicus of which he seems to have admitted only the least important part, that is, the diurnal motion of the earth, and not the much more important annual one. Gilbert, it is true, does not reject this latter: he simply ignores it, whereas he devotes a number of very eloquent pages to the defence and explanation (on the basis of his magnetic philosophy) of the daily rotation of the earth on its axis and to the refutation of the Aristotelian and Ptolemaic conception of the motion of the celestial sphere, and also to the denial of its very existence.

As to this latter point, we must not forget, however, that the solid orbs of classical — and Copernican — astronomy had, in the meantime, been "destroyed" by Tycho Brahe. Gilbert, therefore, in contradistinction to Copernicus himself, can so much more easily dispense with the perfectly useless sphere of the fixed stars, as he does not have to admit the existence of the potentially useful planetary ones. Thus he tells us:

> But in the first place, it is not likely that the highest heaven and all these visible splendours of the fixed stars are impelled along that most rapid and useless course. Besides, who is the Master who has ever made out that the stars which we call fixed are in one and the same sphere, or has established by any reasoning that there are any real, and, as it were, adamantine spheres? No one has ever proved this as a fact; nor is there a doubt but that just as the planets are at unequal distances from the earth, so are those vast and multitudinous lights separated from the earth by varying and very remote altitudes; they are not set in any sphaerick frame of firmament (as is feigned), nor in any vaulted body; accordingly the intervals of some are, from their unfathomable distance, matter of opinion rather than of verification; others do much exceed them and

are very far remote, and these being located in the heaven at varying distances, either in the thinnest aether, or in that most subtle quintessence, or in the void; how are they to remain in their position during such a mighty swirl of the vast orbe of such uncertain substance . . .

Astronomers have observed 1022 stars; besides these innumerable other stars appear minute to our senses; as regards still others, our sight grows dim, and they are hardly discernible save by the keenest eye; nor is there any possessing the best power of vision that will not, while the moon is below the horizon and the atmosphere is clear, feel that there are many more, indeterminable and vacillating by reason of their faint light, obscured because of the distance.

How immeasurable then must be the space which stretches to those remotest of the fixed stars! How vast and immense the depth of that imaginary sphere! How far removed from the earth must the most widely separated stars be and at a distance transcending all sight, all skill and thought! How monstrous then such a motion would be!

It is evident then that all the heavenly bodies, set as if in a destined place, are there formed unto spheres, that they tend to their own centres and that round them there is a confluence of all their parts. And if they have motion that motion will rather be that of each round its own centre, as that of the earth is, or a forward movement of the centre in an orbit as that of the Moon.

But there can be no movement of infinity and of an infinite body, and therefore no diurnal revolution of the *Primum Mobile*.[47]

III. The New Astronomy Against the New Metaphysics

Johannes Kepler's Rejection of Infinity

The conception of the infinity of the universe is, of course, a purely metaphysical doctrine that may well — as it did — form the basis of empirical science; it can never be based on empiricism. This was very well understood by Kepler who rejects it therefore — and this is very interesting and instructive — not only for metaphysical, but also for purely scientific reasons; who even, in anticipation of some present-day epistemologies, declares it scientifically meaningless.[1]

As for the metaphysical reasons for which Kepler denies the infinity of the universe, they are derived chiefly from his religious beliefs. Indeed, Kepler, a devout though somewhat heretical Christian, sees in the world an expression of God, symbolizing the Trinity[2] and embodying in its structure a mathematical order and harmony. Order and harmony that cannot be found in the infinite and therefore perfectly formless — or uniform — universe of Bruno.

Yet it is not this conception of God's creative action,

but a conception of astronomical science, as based upon, and limited by, the phenomena that Kepler opposes to Bruno and to those who share his views. Thus, discussing the interpretation to be given to the appearance of a new star in the foot of the *Serpentarius*, Kepler raises the question whether this amazing and striking phenomenon does not imply the infinity of the universe. He does not think so, yet he knows, and tells us that,[3]

> . . . there is a sect of philosophers, who (to quote the judgment of Aristotle, unmerited however, about the doctrine of the Pythagoreans lately revived by Copernicus) do not start their ratiocinations with sense-perception or accommodate the causes of the things to experience: but who immediately and as if inspired (by some kind of enthusiasm) conceive and develop in their heads a certain opinion about the constitution of the world; once they have embraced it, they stick to it; and they drag in by the hair [things] which occur and are experienced every day in order to accommodate them to their axioms. These people want this new star and all others of its kind to descend little by little from the depths of nature, which, they assert, extend to an infinite altitude, until according to the laws of optics it becomes very large and attracts the eyes of men; then it goes back to an infinite altitude and every day [becomes] so much smaller as it moves higher.
>
> Those who hold this opinion consider that the nature of the skies conforms to the law of the circle; therefore the descent is bound to engender the opposite ascent, as is the case with wheels.
>
> But they can easily be refuted; they indulge indeed in their vision, born within them, with eyes closed, and their ideas and opinions are not received by them [from valid experience] but produced by themselves.

This general criticism may be sufficient. Yet Kepler does not content himself with it and continues: [4]

> We shall show them that by admitting the infinity of the fixed stars they become involved in inextricable labyrinths.
>
> Furthermore we shall, if possible, take this immensity away from them: then, indeed, the assertion will fall of itself.

Kepler knows quite well that this particular opinion concerning the infinity of the world goes back to the ancient heathen philosophers, criticized — rightly, according to him — by Aristotle.[5]

> This particular school of the ancient heathen philosophers is chiefly refuted by the argument by which Aristotle demonstrated the finitude of the world from motion.

As for the moderns, he tells us that the infinity of the world [6]

> . . . was defended by the unfortunate Jord. Bruno. It was also asserted in a by no means obscure way, though he expressed himself as if he doubted it, by William Gilbert in the otherwise most admirable book *De magnete*. Gilbert's religious feeling was so strong that, according to him, the infinite power of God could be understood in no other way than by attributing to Him the creation of an infinite world. But Bruno made the world so infinite that [he posits] as many worlds as there are fixed stars. And he made this our region of the movable [planets] one of the innumerable worlds scarcely distinct from the others which surround it; so that to somebody on the Dog Star (as, for instance, one of the Cynocephals of Lucian) the world would appear from there just as the fixed stars appear to us from our world. Thus according to them, the new star was a new world.

Neither Bruno's enthusiasm for the infinity of the universe, nor even Gilbert's desire to enhance God's infinite power, is shared by Kepler. Quite the contrary, he feels that [7]

> This very cogitation carries with it I don't know what secret, hidden horror; indeed one finds oneself wandering in this immensity, to which are denied limits and center and therefore also all determinate places.

From the purely religious point of view, it would be sufficient, perhaps, to make an appeal to the authority of Moses. Yet the question we are discussing is not a dogmatic one; it has to be dealt with not by recourse to revelation, but by scientific reasoning,[8]

> But because this sect misuses the authority of Copernicus as well as that of astronomy in general, which prove — particularly the Copernican one — that the fixed stars are at an incredible altitude: well then we will seek the remedy in astronomy itself.

Thus by the same means which seem to those philosophers to enable them to break out of the limits of the world into the immensity of infinite space, we will bring them back. "It is not good for the wanderer to stray in that infinity."

Kepler's refutation of the infinitist conception of the universe may appear to the modern reader unconvincing and even illogical. Yet, as a matter of fact, it is a perfectly consistent and very well-reasoned argument. It is based on two premises, which, by the way, Kepler shared with his opponents. The first one is a direct consequence of the principle of sufficient reason and consists in admitting that, if the world has no limits and no particular, determined, structure, that is, if the world-space is infinite and uniform, then the distribution of the fixed stars in this universe must be uniform, too.[9] The second premise concerns the science of astronomy as such. It postulates its empirical character; it tells us that astronomy, as such, has to deal with observable data, that is, with the appearances (*φαινόμενα*); that it has to adapt its hypotheses — for instance, the hypotheses concerning the celestial motions — to these appearances, and that it has no right to transcend them by positing the existence of things that are either incompatible with them, or, even worse, of things that do not and cannot "appear." Now these "appearances" — we must not forget that Kepler is writing in 1606, that is, before the enlargement of the observable data by the discovery and the use of the telescope — are the aspects of the world that we *see*. Astronomy therefore is closely related to sight, that is, to optics. It cannot admit things that contradict optical laws.

Let us now turn back to Kepler: [10]

> To begin with, it can most certainly be learnt from astronomy that the region of the fixed stars is limited downwards; . . . moreover it is not true . . . that this inferior world with its sun differs in no way in its aspect from any one of the fixed stars; that is, [that there is no difference] of one region or place from another.
>
> For, be it admitted as a principle that the fixed stars extend themselves *in infinitum*. Nevertheless it is a fact that in their innermost bosom there will be an immense cavity, distinct and different in its proportions from the spaces that are between the fixed stars. So that if it occurred

> to somebody to examine only this cavity, even [if he were] ignorant of the eight small bodies which fly around the centrum of this space at a very small distance from it, and did not know what they are, or how many; nevertheless from the sole comparison of this void with the surrounding spherical region, fitted with stars, he certainly would be obliged to conclude that this is a certain particular place and the main cavity of the world. Indeed, let us take, for instance, three stars of the second magnitude in the belt of Orion, distant from each other by 81′, being, each one, of at least 2 minutes in diameter. Thus, if they were placed on the same spherical surface of which we are the center, the eye located on one of them would see the other as having the angular magnitude of about $2\frac{3}{4}^\circ$; [a magnitude] that for us on the earth would not be occupied by five suns placed in line and touching each other. And yet these fixed stars are by no means those that are the nearest to each other; for there are innumerable smaller ones that are interspersed [between them]. Thus if somebody were placed in this belt of Orion, having our sun and the center of the world above him, he would see, first, on the horizon, a kind of unbroken sea of immense stars *quasi*-touching each other, at least to the sight; and from there, the more he raised his eyes, the fewer stars would he see; moreover, the stars will no longer be in contact, but will gradually [appear to be] more rare and more dispersed; and looking straight upward he will see the same [stars] as we see, but twice as small and twice as near to each other.

Kepler's reasoning is, of course, erroneous. But only because the data available to him are faulty. In itself it is quite correct. Indeed, if we assume that the fixed stars, or at least the equally bright ones, are at an approximately equal distance from us, if we assume, moreover, that their visible diameter corresponds to their *real* one, we are bound to admit that the two big stars in the belt of Orion, separated by the angular distance of 81′, will be *seen* from each other as covering more surface of the sky than five suns put together; the same will be the case for a great number of the other fixed stars, and therefore the visible aspect of the sky will be, for the observer placed on the fixed stars, quite different from its aspect for us. This implies, of course, a variation in the pattern of the real distribution of the fixed stars in space, that is, the negation of the homogeneity and the uniformity of the universe. Once more, let us not forget that Kepler wrote before the invention of the telescope and did not — and even could not — know that the visible diameter of the fixed stars is a pure optical illusion that gives us no information about their size and distance. Not knowing it, he was entitled to conclude: [11]

> For us the fact of the sky is quite different. Indeed we see everywhere stars of different magnitude, and [we see them] also equally distributed everywhere. Thus around Orion and the Twins we see many of them big and closely packed together: the eye of the Bull, the Capella, the heads of the Twins, the Dog, the shoulders, the belt and the foot of Orion. And in the opposite part of the sky there are equally large ones: the Lyre, the Eagle, the heart and the brow of the Scorpion, the Serpentarius, the arms of the Balance; and before them Arcturus; the head of the Virgin; also after them the last star of the Water Bearer and so on.

I have just pointed out that Kepler's discussion of the astronomical data that enabled him to assert the particular, unique structure of our site in the world-space was based on the assumption of the equidistance — from us

— of the fixed stars. Couldn't this conclusion be avoided if we admitted that the stars are so far away from us — and therefore from each other — that, seen from each other, they will not appear as big as we have calculated? Or couldn't we go even farther and admit that our fundamental assumption could, possibly, be incorrect and that stars which *appear* to be near each other could, in point of fact, be separated by an enormous distance, the one being near us and the other exceedingly far away? As we shall see, even if it were so, it would not change the fundamental fact of the singularity of our world-space. But the objection has to be dealt with. Kepler, therefore, proceeds: [12]

> When, some time ago, I advanced these views [just developed] some people, to try me, vigorously defended the cause of infinity, which they had taken from the above-mentioned philosophers. They asserted that, granting infinity, it was easy for them to separate the pairs of fixed stars (which we on the earth perceive as being very near each other) by as great a distance as that which separates us from them. Yet this is impossible. Even admitting that you can arbitrarily elevate [13] the double fixed stars [that are] equally distant from the center of the world, it must be remembered that, if we elevate the fixed stars, the void which is in the middle, and also the circular envelope of the fixed stars, increase at the same time. Indeed, [these people] assume thoughtlessly that, the fixed stars being elevated, the void will remain the same.

As it will not, the singular character of our site will be maintained.[14]

> But what, they say, if, of the two stars of the belt of Orion, we assume one to remain in its sphere, because the

theory of parallaxes does not admit an inferior position,[15] and the other to be higher by an infinite distance? Shall we not, in this way, obtain that, seen from each other, they appear as small as they appear to us? and that there will be a distance between them, void of stars, equal to the distances between us and them?

I answer that, perhaps, one could use this method if there were only two stars, or only a few of them, and if they were not dispersed and disseminated in a circle. Indeed, you either alternately remove the stars to a greater distance and let them stay where they are or [you remove them] all together. If alternately, you do not solve the problem, though you decrease somewhat the difficulty. For, concerning those that will remain near, the affirmation [made by us] will still be just as valid. The pairs of stars will be nearer to each other than to the sun, and their diameters, as seen from each other, larger [than they are as seen by us]. But those that are removed higher will, of course, be more distant [from each other], yet nevertheless they will be comparatively large [as seen from each other]. And I would even easily concede, without endangering my cause, that all the fixed stars are of the same magnitude; and that those which to us appear large are near to us, and those [which appear small] are so much farther. As sings Manilius: [15a] 'Not because less bright, but because they are removed to a greater altitude.'

I say: I will concede not: I will assert. For it is just as easy to believe that [the stars] differ really in brightness, in color and also in magnitude. And it is possible that both [opinions] are true, as is the case with the planets, of which some are really larger than others, whereas some others only appear to be larger though in themselves they are smaller, namely because they are nearer to us.

The consequences of these hypotheses will be seen later.

For the moment we have to discuss the implications for the φαινόμενα of a really uniform distribution of the fixed stars in the world-space, that is, of a distribution according to which they would be separated from each other by equal distances, namely by the same distance that separates them from us.[16]

> But let us pass to the other member [of the argument], and say what would result if all the stars were separated from each other by the same distance, in such a way that the nearest ones would retain the propinquity which astronomy imposes as a limit to all [stars], not allowing any one to be nearer, and all the others would be elevated in respect to it, and removed to an altitude equal to the distance of the nearest one to us.
>
> As a matter of fact nothing will result from all this. It will never be the case that the [starry heavens] would appear to those whom we may imagine observing them from these stars as they appear to us. From which it follows that this place, in which we are, will always have a certain peculiarity that cannot be attributed to any other place in all this infinity.

Once more, in order to understand Kepler's reasoning, we have to remember that we are not discussing the abstract possibility of a certain distribution of stars in the world-space, but the concrete distribution of stars corresponding to the *appearance* of the sky; that is, we are dealing with the distribution of visible stars, of those that we actually *see*. It is *their* distance from us that is in question, and it is for *them* that the possibility of a uniform distribution, which would place most of them at very great, and regularly increasing, distances from us, is denied.[17]

For, if the state of affairs were such as has been said, it is certain that those stars that are two, three, a hundred times higher will also be two, three, a hundred times larger. Indeed, let a star be as elevated as you wish, you will never obtain that it would be seen by us as having a diameter of two minutes.[18] Thus the diameter will always be two thousandth, one thousandth, or so of the distance from us; but this diameter will be a much larger part of the mutual distances between two fixed stars (since these distances are much smaller than their distance from us). And though from a star near us the face of the sky will appear nearly the same as it does to us; yet from the other stars the aspect of the world will be different, and all the more different in that they are farther. Indeed, if the intervals of the pairs of stars (which to us appear as nearest each other) remain constant, their aspects [dimensions], as seen from each other, will increase [with their distance from us]. For the more you remove the stars to an infinite altitude, the more monstrous you imagine their dimensions, such as are not seen from this place of our world.

An observer starting from the earth and moving upwards to the outer spaces would, therefore, find the " appearance " of the world constantly changing, and the fixed stars always increasing in their real as well as visible dimensions. Besides,[19]

The same must be said concerning the space that for such a traveler increases continuously, every time he transfers the stars from one order to the next and moves them higher. You may say that he is building the shell of a snail, which becomes ever wider towards the exterior.

You cannot, indeed, separate the stars [by moving them] downward; the theory of the parallaxes does not allow it because it puts a certain limit to the approximation; you cannot separate them sideways, as they possess already their places determined by sight; it remains thus to separate the stars by moving them upward, but in this case the space that surrounds us and in which are found no stars whatever except the eight small globes in the very centre of this void, grows at the same time.

Thus it is obvious that we may assume the world to be as large as we like; still the disposition of the fixed stars *as seen by us* will be such that this our place will appear as possessing a certain particularity and as having a certain manifest property (the absence of fixed stars in the vast void) by which it is distinct from all other places.

Kepler is perfectly right. We can make the world as big as we wish, and yet, if we have to restrict its contents to the *visible* stars, which moreover appear to us as finite, measurable bodies — not points of light — we will never be able to assign to them a uniform distribution that would " save " the phenomena. Our world will always be distinguished by a particular structure.[20]

It is certain that, on the inside, toward the sun and the planets, the world is finite and, so to say, excavated. What remains belongs to metaphysics. For, if there is such a place [as our world] in this infinite body, then this place will be in the center of the whole body. But the fixed stars which surround it will not, in respect to it, be in a position similar [to that of our sun] as they should be if there were everywhere worlds similar to ours. But they will form a closed sphere around this [void]. This is most obvious in the case of the Milky Way which passes through [the heavenly sphere] in an uninterrupted circle, holding us in the middle. Thus both the Milky Way and the fixed stars play the role of extremities. They limit this our space, and in turn are limited on the exterior. Is it, indeed, credible that, having a limit on this side, they extend on the other side to infinity? How can we find in infinity a centrum which, in infinity, is everywhere? For every point taken in the infinity is equally, that is, infinitely, separated from the extremities which are infinitely distant. From which it would result that the same [place] would be the center and would not be [the center], and many other contradictory things, which most correctly will be avoided by the one who, as he found the sky of the fixed stars limited from inside, also limits it on the outside.

Yet, can we not assume that the region of the fixed stars is boundless and that stars follow upon stars, though some, or even most of them, are so far away that we do not see them? Assuredly we can. But it will be a purely gratuitous assumption, not based on experience, that is, on sight. These invisible stars are not an object of astronomy and their existence cannot in any way be demonstrated.

In any case there cannot be stars — especially visible ones — at an actually infinite distance from us. Indeed, they should necessarily be infinitely large. And an infinitely large body is utterly impossible because it is contradictory.

Once more Kepler is right. A visible star cannot be at an infinite distance; nor, by the way, can an invisible one: [21]

If there were an infinite altitude of the sphere of the fixed stars, that is, if some fixed stars were infinitely high, they would also be in themselves of an infinite corporeal bulk.

> Imagine, indeed, a star, seen under a certain angle, for instance, 4′; the amplitude of such a body is always a thousandth part of its distance, as we know from geometry. Consequently if the distance is infinite, the diameter of the star will be the thousandth part of the infinite. But all the aliquot parts of the infinite are infinite. Yet at the same time it will be finite, because it has a form: all form is circumscribed by certain bounds, that is, [all form] is finite or limited. But we have given it a form when we have posited it as visible under a certain angle.

The impossibility of a visible star's being at an infinite distance thus demonstrated, there remains the case of an invisible one.[22]

> But what, you will ask, if it were so small as not to be seen? I answer that the result is the same. It is necessary, indeed, that it occupy an aliquot part of the circumference that passes through it. But a circumference of which the diameter is infinite is itself infinite. Thus it follows that no star, either visible, or having vanished because of its smallness, is separated from us by an infinite distance.

It remains only to ask ourselves whether an infinite space without stars can be posited. Kepler replies that such an assertion is utterly meaningless, since wherever you put a star you will be at a finite distance (from the earth) and if you go beyond, you cannot speak of a distance.[23]

> Finally, even if you extend the place without stars to infinity, it is certain that wherever you put a star into it, you will have a finite interval and a finite circumference determined by the star; thus, those who say that the sphere of the fixed stars is infinite commit a contradiction *in ad-*

jecto. In truth, an infinite body cannot be comprehended by thought. For the concepts of the mind concerning the infinite are either about the meaning of the term "infinite," or about something which exceeds all conceived numerical, visual, tactual measure: that is, something which is not infinite *in actu*, as an infinite measure can never be thought of.

Kepler, once more, is perfectly, or at least partially, right. It is quite certain that wherever you put a star you will find yourself at a finite distance from your starting point, as well as from all other stars in the universe. A really infinite distance between two bodies is unthinkable, just as an infinite integer is unthinkable: all integers that we can reach by counting (or any other arithmetical operation) are necessarily finite. Yet it is perhaps too rash to conclude therefore that we have no concept of the infinite: does it not mean precisely — as Kepler tells us himself — that it is what is "beyond" all number and all measure?

Furthermore, just as in spite of — or because of — the finiteness of all numbers we can go on counting without end, can we not also go on putting stars in space, all, of course, at finite distances, without ever coming to an end? Certainly we can, provided we abandon Kepler's empirical, that is, Aristotelian or semi-Aristotelian, epistemology which precludes this operation, and replace it by another: an *a priori* Platonic or at least semi-Platonic one.

In my analysis of Kepler's objections to the infinity of the world I have pointed out that they were formulated several years *before* the great astronomical (telescopical) discoveries of Galileo. These discoveries, which so tremendously enlarged the field of observable stars and so deeply modified the aspect of the celestial vault, discoveries which Kepler accepted and defended with joy, and which he supported not only with the weight of his undisputed authority but also by establishing the theory of the instrument — the telescope — used by Galileo, obliged him, of course, to modify some of the views he had expressed in his treatise on the new star. However, and this seems to me extremely interesting and significant, they did not lead him to the acceptance of the infinitist cosmology. On the contrary, they seemed to him to confirm his own finitistic world-view and to bring new data in favor of the unicity of the solar system and of the essential distinction of our moving world and the motionless congeries of the fixed stars.

Thus in his famous *Dissertatio cum nuntio sidereo* he tells us that at first, before having in hand the publication of Galileo, he was somewhat disturbed by the conflicting reports about the latter's discoveries, namely, whether the new stars were new planets moving around the sun, new "moons" accompanying the solar planets, or, as his friend Mattheus Wackher believed, planets revolving around some fixed stars: a strong argument in favor of Bruno's conception of the uniformity of the world. In this case, indeed,[24]

> . . . nothing could prevent us from believing that numberless others would be discovered later on, and that either this our world were infinite as Melissos and the author of magnetic philosophy, William Gilbert, held, or that there was an infinity of worlds and earths (besides this one) as was believed by Democritus and Leucippus and, among the moderns, by Bruno, Brutus, Wacherus and, possibly also, by Galileo.

The perusal of the *Nuntius* tranquillized Kepler. The new stars were not planets: they were moons, Jupiter's moons. Now, if the discovery of *planets* — whether revolving around fixed stars or around the sun — would have been extremely disagreeable for Kepler, the discovery of new *moons* did not affect him at all. Why, indeed, should the earth be the only planet to possess a moon? Why should the other ones not be similarly endowed with satellites? There is no reason why the earth should have this privilege. Nay, Kepler thinks that there are good reasons why all the planets — with the exception perhaps of Mercury, too near the sun to need one — should be surrounded with moons.

It could be said, of course, that the earth has a moon because it is inhabited. Thus, if the planets had moons, they should be inhabited too. And why shouldn't they be? There is, according to Kepler — who, *for our world,* accepts the teachings of Nicholas of Cusa and Bruno — no reason to deny this possibility.

As for the other discoveries of Galileo, namely, those concerning the fixed stars, Kepler points out that they enhance the difference between the stars and the planets. Whereas the latter are strongly magnified by the telescope and appear as well-defined discs, the former hardly increase their dimensions for, seen through the telescope, they are deprived of the luminous haze that surrounds them,[25] a fact of tremendous importance because it shows that this haze belongs not to the seen stars but to the seeing eye, in other words, that it is not an objective but a subjective phenomenon and that, whereas the visible dimensions of the planets have a determinate relation to their real ones, this is not the case for the fixed stars.

Thus we can calculate the dimensions of the planets, but we cannot do it, at least not as easily, for the fixed stars.

The explanation of this fact is easy: whereas the planets shine by the reflected light of the sun, the fixed stars shine by their own, like the sun. But if so, are they not really suns as Bruno has asserted? By no means. The very number of the new stars discovered by Galileo proves that the fixed stars, generally speaking, are much smaller than the sun, and that there is in the whole world not a single one which in dimensions, as well as in luminosity, can be equal to our sun. Indeed, if our sun were not incommensurably brighter than the fixed stars, or these so much less bright than it, the celestial vault would be as luminous as the sun.

The very existence of a tremendous number of fixed stars which *we* do not see, but which observers placed upon one of them would, is a proof, according to Kepler, that his fundamental objection to the infinitist cosmology, namely, that for no observer in the world would the aspect of the sky be the same as it is for us, is even better grounded in the facts than he had imagined. Thus the conclusion formerly drawn from the analysis of the phenomena accessible to the unassisted eye finds itself confirmed by the adjunction to them of the phenomena revealed by the telescope: our moving world, with its sun and planets, is not one of many, but a unique world, placed in a unique void, surrounded by a unique conglomeration of innumerable fixed — in the full sense of the term — stars.

Kepler thus maintains his position. Of the two possible interpretations of the telescopic discoveries of Galileo, that the new (fixed) stars are not seen by the unassisted eye because they are too far, and that they are not seen because they are too small, he resolutely adopts the second.

He is wrong, of course; and yet, from the point of view of pure empiricism, he is blameless because there are, for him, on the one hand, no means of determining the intervals that separate us from the stars and no reason therefore to assume that they are not very different in size; all the more so as there are, on the other hand, some examples — the " Medicean " planets, in fact — of celestial objects imperceptible because they are too small to be seen.

Let us turn now to the *Epitome astronomiae Copernicanae*, the last, and the most mature, great work of Kepler. We shall find the rejection of the infinity of the world presented just as vigorously, or perhaps even more vigorously, than ever before. To the question [26]

What is to be held concerning the shape of the sky?

the reply is given:

> Though we cannot perceive with our eyes the matter of the etheric aura, there is nothing, however, to prevent us from believing that it is spread through the whole amplitude of the world on all sides surrounding the elementary sphere. That the army of the stars completely encircles the earth and thus forms a certain *quasi*-circular vault is clear from the fact that, while the earth is round, men, wherever they go, see the stars above their heads, as we do.

Thus if we turned around the earth, or if the earth turned around with us, we would see the whole troop of the stars arranged in a closed circuit. But that is not an answer to the question asked, as nobody doubts that the earth is surrounded by stars. What we have to find out is something quite different, namely, whether this *quasi* vault is more than a simple appearance, that is whether [27]

> *the centres of the stars are placed on the same spherical surface.*

At this stage of the discussion Kepler does not want to commit himself. Thus he gives a rather cautious answer:

> This is rather uncertain. As some of them are small, and others big, it is not impossible that the small ones appear such because they are far away in the high ether, and the large [do so] because they are nearer to us. Nor is it absurd that two fixed [stars] of different apparent magnitude be distant from us by the same interval.
>
> As for the planets, it is certain that they are not in the same spherical surface as the fixed stars; indeed they eclipse the fixed stars but are not eclipsed by these.

But in this case, that is, if we can neither determine the intervals that separate us from the fixed stars nor decide whether their apparent magnitude is a function of their real size or only of their distance, why should we not admit that their " region " is unlimited or infinite? Indeed,[28]

> *If there is no more certain knowledge concerning the fixed stars, it would seem that their region is infinite; nor will this our sun be anything other than one of the fixed stars, larger and better seen by us, because* [*it is*] *nearer to us than the fixed stars; and in this case around any one of the fixed stars there may be such a world as there is around us; or, which is exactly the same, among the innumerable places in that infinite assembly of the fixed stars*

our world with its sun will be one [*place*] *in no way different from other places around other fixed stars, as* [*represented*] *by the adjoined figure M.*

The supposition seems reasonable or, at least, admissible. Yet Kepler rejects it, and does so for the same reasons he had twelve years before: from the hypothesis of infinity, that is, of a uniform distribution of the fixed stars in space, would follow an aspect of the sky that is not in accordance with the phenomena. For Kepler, indeed, the infinity of the world necessarily implies a perfect uniformity of its structure and contents. An irregular, irrational scattering of fixed stars in space is unthinkable; finite or infinite, the world must embody a geometrical pattern. But whereas for a finite world it is reasonable to choose a particular pattern, the principle of sufficient reason prevents the geometrically minded God of Kepler from doing it in an infinite one. As already explained by Bruno, there is no reason (or even possibility) for God to make a distinction between the "places" of a perfectly homogeneous space, and to treat them in a different way. Kepler thus states: [29]

> This [the infinity of the world] indeed [was asserted] by Bruno and some others. But [even] if the centers of the fixed stars are not on the same spherical surface, it does not follow that the region in which they are dispersed is everywhere similar to itself.
>
> As a matter of fact, in the midst of it [the region of the fixed stars] there is assuredly a certain immense void, a hollow cavity, surrounded in close order by the fixed stars, enclosed and circumscribed as by a wall or vault; it is in the bosom of this immense cavity that our earth with the sun and the moving stars [planets] is situated.

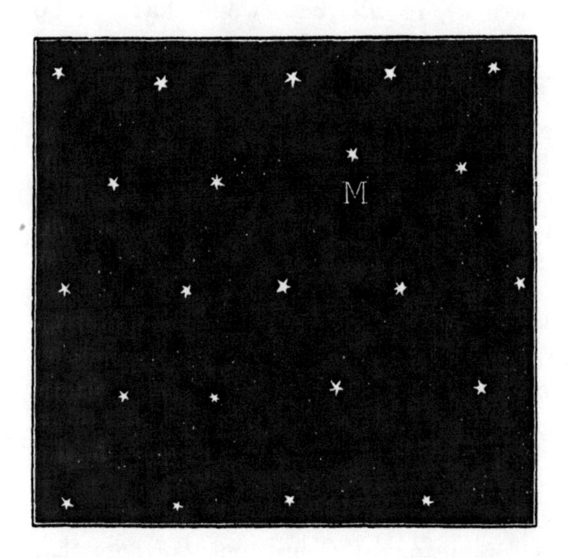

FIGURE 3

The figure M *of Kepler*

(from the *Epitome astronomiae Copernicanae,* 1618)

In order to demonstrate this assertion, Kepler gives us a detailed description of the aspect that the sky would have in the case of a uniform distribution of the fixed stars (which, moreover, *in this case* would have to be assumed as being, all of them, of the same size), and opposes this hypothetical picture to the actual one.[30]

> If the region of the fixed stars were everywhere similarly set with stars, even in the vicinity of our movable world, so that the region of our world and of our sun had no peculiar outline compared to the other regions, then only a few enormous fixed stars would be seen by us, and not more than twelve (the number of the angles of the icosahedron) could be at the same distance from us and of the same [visible] magnitude; the following ones would be scarcely more numerous, yet they would be twice as distant as the nearest ones; the next higher would be three times as far, and so on, always increasing their distance [in the same manner].
>
> But as the biggest of all appear so small that they can hardly be noted or measured by instruments, those that would be two or three times farther off, if we assume them to be of the same true magnitude, would appear two or three times smaller. Accordingly we should quickly arrive at those which would be completely imperceptible. Thus very few stars would be seen, and they would be very different from each other.
>
> But what is seen by us in fact is quite different. We see, indeed, fixed stars of the same apparent magnitude packed together in a very great number. The Greek astronomers counted a thousand of the biggest, and the Hebrews eleven thousand; nor is the difference of their apparent magnitudes very great. All these stars being equal to the sight, it is not reasonable that they should be at very unequal distances from us.

Thus, as the general appearance of the fixed stars is everywhere nearly the same in respect to their number and magnitude, the visible sky is also everywhere raised above us by nearly the same distance. There is therefore an immense cavity in the midst of the region of the fixed stars, a visible conglomeration of fixed stars around it, in which enclosure we are.

In the belt of Orion there are three big stars which are distant from each other by an interval of 83′; let us suppose the visible semidiameter of each to be only of one minute; accordingly it will appear to the sight as being of 83′, that is, nearly threc times the breadth of the sun, and as for the surface, it would be eight times larger than the sun itself. Consequently the appearance of the fixed stars as seen from each other is not the same as it is from our world, and accordingly we are farther away from the fixed stars than the neighbouring fixed stars are from each other.

As we see, the telescope did not change the pattern of Keplerian reasoning: it only made him diminish somewhat the visible dimensions of the fixed stars. And, of course, as long as this visible dimension is not completely removed from the objective sphere to the subjective one, Kepler's deduction can be upheld.

Yet, it may be objected, its second premise, that of the uniform size of the fixed stars, is gratuitous. It seems that,[31]

The strength of this argument can be weakened by assuming that the stars are so much larger as they are higher [farther] from the earth. For, if among the so numerous stars that are seen under nearly the same angle, some were assumed to have small bodies, and others enormous ones, it would follow that the former are near us and the latter exceedingly far; and thus, in this case, stars which are seen by us as very near [*to each other*] *could in point of fact be very distant.*

This is a possible assumption, but, as we know, a rather improbable one, since it would imply an extremely unlikely star distribution, a distribution, moreover, completely incompatible with our fundamental assumption of a homogeneous, uniform universe: [32]

> In this case, this region would be conspicuous if not by its vacuity then by the smallness of the stars in the neighbourhood of our moving world, and thus the very minuteness of the stars would present a kind of void, whereas the increasing magnitude of the stars on the exterior would play the role of the vault. In the universe there would be less stellar matter in this cavity in which our moving world is located, and more matter in the circumference which contains and limits it. Thus it would still remain true that this place is singular and notable compared to all the remaining parts of the region of the fixed stars.
>
> Moreover, it is more probable that those [stars] that are nearly of the same sensible magnitude are separated from us by nearly the same distance, and that a kind of hollow sphere is formed by the packing closely together of so many stars.

The arguments already developed are more than sufficient to enable us to maintain the unicity of this our moving and sun-centered world, and to oppose it to the realm of the fixed stars. We can, however, supplement them by more direct ones, and show that the phenomena clearly point out our (the solar system's) central position in the midst of the peripheral accumulation of stars. The appearance of the Milky Way — in spite of its resolution by Galileo into an innumerable multitude of stars — still seems to Kepler to preclude any other conclusion. Thus, elaborating the demonstration outlined in the *De stella nova*, Kepler continues: [33]

> *Do you have any other argument demonstrating that this place in the midst of which are the earth and the planets is particularly distinguished in respect to all other places in the region of the fixed stars?*
>
> The way called by the Greeks the Milky Way and by us the Road of St. Jacob is spread around in the middle of the orb of the fixed stars (as the orb appears to us), dividing it into two apparent hemispheres; and though this circle is of unequal breadth, still it is, all around, not very dissimilar to itself. Thus the Milky Way conspicuously determines the place of the earth and of the moving world in relation to all other places in the region of the fixed stars.
>
> For if we assume that the earth is on one side of the semidiameter of the Milky Way, then this Milky Way would appear to it [the earth] as a small circle or small ellipse . . . it would be visible at one glance, whereas now not more than half of it can be seen at any moment. On the other hand, if we assumed that the earth were indeed in the plane of the Milky Way, but in the vicinity of its very circumference: then this part of the Milky Way would appear enormous, and the opposite part, narrow.
>
> Thus the sphere of the fixed stars is limited downwards, towards us, not only by the stellar orb but also by the circle of the Milky Way.

Still, in spite of being thus limited " downwards," the sphere of the fixed stars could nevertheless extend indefinitely " upwards "; the walls of the world-bubble could be indefinitely, or infinitely, thick. Once more we see

Kepler reject this supposition as groundless and perfectly unscientific. Astronomy, indeed, is an empirical science. Its field is coextensive with that of observable data. Astronomy has nothing to say about things that are not, and cannot, be seen.[34]

> *But then is not the region of the fixed stars infinite upwards?* Here astronomy makes no judgment, because in such an altitude it is deprived of the sense of seeing. Astronomy teaches only this: as far as the stars, even the least ones, are seen, space is finite.

Kepler does not mention Galileo in this discussion, and we can understand why: the telescope does not change the situation. It allows us to see more stars than we did before its invention; it enables us to transcend the *factual* limitation of our sense of seeing; but it does not remove its essential structure. With as without the telescope, things at an infinite distance cannot be seen. The optical world is finite.

Thus to the question: [35]

> *But is it not possible for some of the visible stars to be separated from us by an infinite distance?*

Kepler replies:

> No; because everything that is seen, is seen by its extremities. Consequently a visible star has limits all around. But if the star receded to a really infinite distance, these limits too would be distant from one another by an infinite space. For everything at once, that is, the whole body of the star, would participate in the infinity of this altitude. Therefore, if the angle of vision remained the same, the diameter of the star, which is the line between its limits, would be increased proportionally to the distance; thus the

> diameter of a [star] twice as distant will be twice as large as the diameter of the nearer one, the diameter of a [star] distant by a finite space will be finite, but when a body is assumed to acquire an infinitely increased distance [its diameter] also becomes infinitely great.
>
> Indeed, to be infinite and to be limited is incompatible, just as it is incompatible to be infinite and to have a certain, that is, determinate, proportion to something finite. Consequently, nothing that is visible is separated from us by an infinite distance.

So much for the *visible* world. But can we not assume that outside and beyond the world, or the part of the world that is seen by us, space, and stars in space, continue to exist without end? It may be meaningless from the point of view of astronomy, it may be metaphysics. . . . But is it a good one? Not according to Kepler, who held that this concept — that of modern science — is bad, as a really infinite number of finite bodies is something unthinkable, even contradictory: [36]

> *But what if there were in reality stars, of finite body, scattered upwards in the infinite spaces,* [*stars*] *which, because of so great a distance, were not seen by us?*
>
> First, if they are not seen, they in no way concern astronomy. Then, if the region of the fixed stars is at all limited, namely downwards, towards our mobile world, why should it lack limits upwards? Third, though it cannot be denied that there can be many stars which, either because of their minuteness or because of their very great distance, are not seen, nevertheless you cannot because of them assert an infinite space. For if they are, individually, of a finite size, they must, all of them, be of a finite number. Otherwise, if they were of an infinite number, then, be they as small as you like, provided they are not infinitely so, they

would be able to constitute one infinite [star] and thus there would be a body, of three dimensions, and nevertheless infinite, which implies a contradiction. For we call infinite what lacks limit and end, and therefore also dimension. Thus all number of things is actually finite for the very reason that it is a number; consequently a finite number of finite bodies does not imply an infinite space, as if engendered by the multiplication of a multitude of finite spaces.

Kepler's objection against infinity is, of course, not new: it is essentially that of Aristotle. Yet it is by no means negligible, and modern science seems rather to have discarded than to have solved the problem.[37] Now, even if we deny that there is an infinite number of stars in space, there still remains, for the infinitist, a last possibility: that of asserting a finite world immersed in an infinite space.[38] Kepler does not accept this, either, and his reasons for rejecting it reveal the ultimate metaphysical background of his thinking: [39]

If you are speaking of void space, that is, of what is nothing, what neither is, nor is created, and cannot oppose a resistance to anything being there, you are dealing with quite another question. It is clear that [this void space], which is obviously nothing, cannot have an actual existence. If, however, space exists because of the bodies located in it [it will not be infinite as] it is already demonstrated that no body that can be located is actually infinite, and that bodies of finite magnitude cannot be infinite in number. It is therefore by no means necessary that space be infinite on account of the bodies located in it. And it is also impossible that between two bodies there be an actually infinite line. For it is incompatible to be infinite and to have limits in the two individual bodies or points that constitute the ends of the line.

Space, void space, is just "nothing," a *non-ens*. Space, as such, neither *is* — how, indeed, could it *be* if it is nothing? — nor has it been created by God, who assuredly has created the world out of nothing, but did not start by creating "nothing."[40] Space exists on account of the bodies; if there were no bodies, there would not be space. And if God should destroy the world, there would be no void space left behind. There would be simply *nothing*, just as there was *nothing* at all before God created the world.

All that is not new, nor specific to Kepler: it is the traditional teaching of Aristotelian scholasticism. Thus we have to admit that Johannes Kepler, the great and truly revolutionary thinker, was, nevertheless, bound by tradition. In his conception of being, of motion, though not of science, Kepler, in the last analysis, remains an Aristotelian.

IV. Things Never Seen Before and Thoughts Never Thought:

THE DISCOVERY OF NEW STARS IN THE WORLD SPACE AND THE MATERIALIZATION OF SPACE

Galileo
& Descartes

I have already mentioned the *Sidereus Nuncius*[1] of Galileo Galilei, a work of which the influence — and the importance — cannot be overestimated, a work which announced a series of discoveries more strange and more significant than any that had ever been made before. Reading it today we can no longer, of course, experience the impact of the unheard-of message; yet we can still feel the excitement and pride glowing beneath the cool and sober wording of Galileo's report: [2]

> In this little treatise I am presenting to all students of nature great things to observe and to consider. Great as much because of their intrinsic excellence as of their absolute novelty, and also on account of the instrument by the

aid of which they have made themselves accessible to our senses.

It is assuredly important to add to the great number of fixed stars that up to now men have been able to see by their natural sight, and to set before the eyes innumerable others which have never been seen before and which surpass the old and previously known [stars] in number more than ten times.

It is most beautiful and most pleasant to the sight to see the body of the moon, distant from us by nearly sixty semidiameters of the earth, as near as if it were at a distance of only two and a half of these measures.

So that

Any one can know with the certainty of sense-perception that the moon is by no means endowed with a smooth and polished surface, but with a rough and uneven one, and, just like the face of the earth itself, is everywhere full of enormous swellings, deep chasms and sinuosities.

Then to have settled disputes about the Galaxy or Milky Way and to have made its essence manifest to the senses, and even more to the intellect, seems by no means a matter to be considered of small importance; in addition to this, to demonstrate directly the substance of those stars which all astronomers up to this time have called *nebulous*, and to demonstrate that it is very different from what has hitherto been believed, will be very pleasant and very beautiful.

But what by far surpasses all admiration, and what in the first place moved me to present it to the attention of astronomers and philosophers, is this: namely, that we have discovered four planets, neither known nor observed by any one before us, which have their periods around a certain big star of the number of the previously known ones, like Venus and Mercury around the sun, which sometimes pre-

> cede it and sometimes follow it, but never depart from it beyond certain limits. All this was discovered and observed a few days ago by means of the *perspicilli* invented by me through God's grace previously illuminating my mind.

To sum up: mountains on the moon, new "planets" in the sky, new fixed stars in tremendous numbers, things that no human eye had ever seen, and no human mind conceived before. And not only this: besides these new, amazing and wholly unexpected and unforeseen facts, there was also the description of an astonishing invention, that of an instrument — the first scientific instrument — the *perspicillum*, which made all these discoveries possible and enabled Galileo to transcend the limitation imposed by nature — or by God — on human senses and human knowledge.[3]

No wonder that the *Message of the Stars* was, at first, received with misgivings and incredulity, and that it played a decisive part in the whole subsequent development of astronomical science, which from now on became so closely linked together with that of its instruments that every progress of the one implied and involved a progress of the other. One could even say that not only astronomy, but science as such, began, with Galileo's invention, a new phase of its development, the phase that we might call the instrumental one.

The *perspicilli* not only increased the number of the fixed, and errant, stars: they changed their aspect. I have already dealt with this effect of the use of the telescope. Yet it is worth while quoting Galileo himself on this subject: [4]

> First of all, this is worthy of consideration, namely that

> stars, as well fixed as errant, when they are seen through the *perspicillum*, are never seen to increase their dimensions in the same proportions in which other objects, and the moon itself, increase in size. Indeed in [the case of] the stars this increase appears much smaller, so that a *perspicillum* which, for instance, is powerful enough to magnify all other objects a hundred times will scarcely render the stars four or five times larger. But the reason for it is this: namely the stars, when seen by our free and natural eyesight, do not present themselves to us with their real and, so to say, naked size, but are surrounded by a certain halo and fringed with sparkling rays, particularly so when the night is already advanced; therefore they appear much larger than [they would] if they were stripped of these adventitious fringes; for the angle of vision is determined not by the primary body of the star, but by the brightness that surrounds it.

According to Galileo, this "adventitious" and "accidental" character of the halo surrounding the stars is clearly demonstrated by the fact that, when they are seen at dawn, stars, even of the first magnitude, appear quite small; and even Venus, if seen by daylight, is hardly larger than a star of the last magnitude. Daylight, so to say, cuts off their luminous fringes; and not only light, but diaphanous clouds or black veils and colored glass have the same effect.[5]

> The *perspicillum* acts in the same way. First it removes from the stars the accidental and adventitious splendours, and [only] then enlarges their true globes (if indeed they are of a round shape), and therefore they appear to be magnified in a smaller proportion [than other objects]. Thus a starlet of the fifth or the sixth magnitude seen through a *perspicillum* is shown only as of the first magnitude.

This, indeed, is extremely important as it destroys the basis of Tycho Brahe's most impressive — for his contemporaries — objection to heliocentric astronomy, according to which the fixed stars — if the Copernican world-system were true — should be as big, nay much bigger, than the whole *orbis magnus* of the annual circuit of the earth. The *perspicillum* reduces their visible diameter from 2 minutes to 5 seconds and thus disposes of the necessity to increase the size of the fixed stars beyond that of the sun. Yet the decrease in size is more than compensated by an increase in number: [6]

> The difference between the appearance of the planets and of the fixed stars seems equally worthy of notice. Planets indeed present their discs perfectly round and exactly delimited, and appear as small moons completely illuminated and globular; but the fixed stars are not seen as bounded by a circular periphery, but like blazes of light, sending out rays on all sides and very sparkling; and with the *perspicillum* they appear to be of the same shape as when viewed by the natural sight, and so much bigger that a starlet of the fifth or sixth magnitude seems to equal the Dog, the largest of all the fixed stars. But below the stars of the sixth magnitude, you will see through the *perspicillum* so numerous a herd of other stars that escape the natural sight as to be almost beyond belief; for you may see more than six other differences of magnitude; of which the largest, those that we may call stars of the seventh magnitude or of the first of the invisible ones, appear with the aid of the *perspicillum* larger and brighter than stars of the second magnitude seen by natural sight. But in order that you may see one or two examples of their nearly inconceivable number, we decided to make out two star-pictures, so that

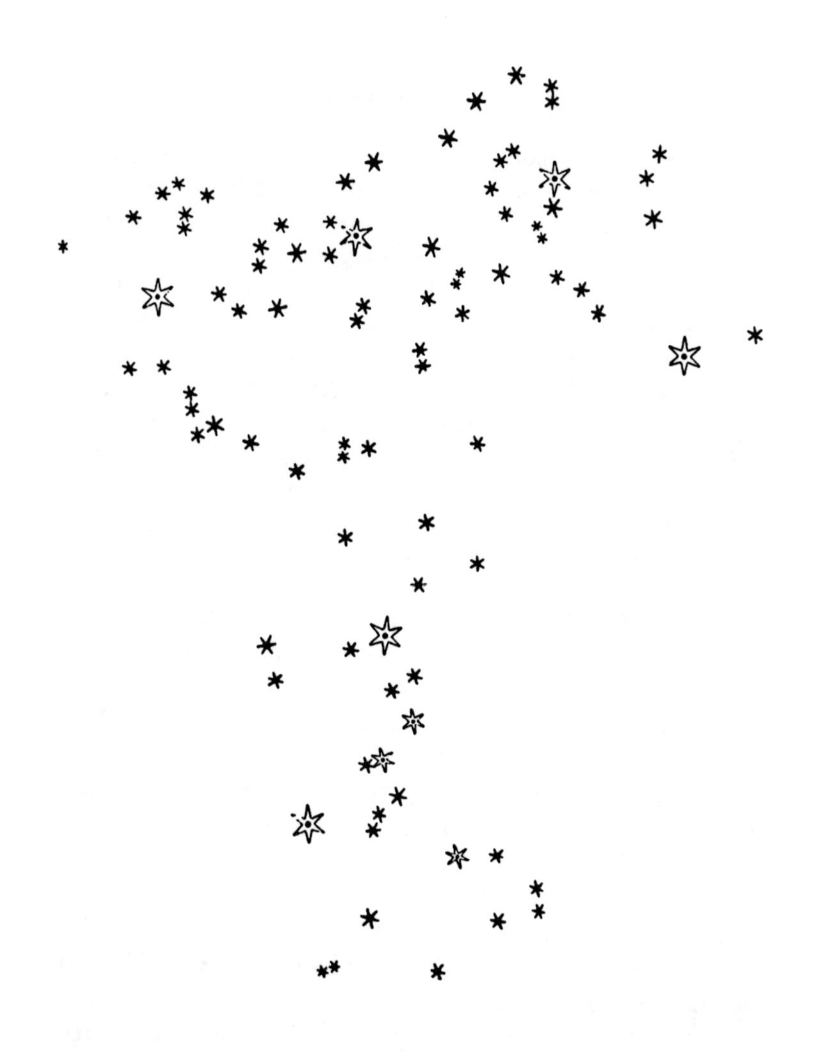

FIGURE 4

Galileo's star-picture of the shield and sword of Orion

(from the *Sidereus Nuncius*, 1610)

> from these examples you may judge about the rest. At first we determined to depict the entire constellation of Orion, but we were overwhelmed by the enormous multitude of stars and by lack of time, and have deferred this attempt to another occasion; for there are adjacent to, or scattered around, the old ones more than five hundred [new ones] within the limits of one or two degrees.
>
> As a second example we have depicted the six stars of Taurus, called the Pleiades (we say six, because the seventh is scarcely ever visible), which are enclosed in the sky within very narrow boundaries, and near which are adjacent more than forty other visible ones, none of which is more than half a degree distant from the aforesaid six.

We have already seen that the invisibility for the human eye of the fixed stars discovered by Galileo, and, accordingly, the role of his *perspicillum* in revealing them, could be interpreted in two different ways: it could be explained by their being (a) too small to be seen, (b) too far away. The *perspicillum* would act in the first case as a kind of celestial microscope, in enlarging, so to say, the stars to perceivable dimensions; in the second it would be a " telescope " and, so to say, bring the stars nearer to us, to a distance at which they become visible. The second interpretation, that which makes visibility a function of the distance, appears to us now to be the only one possible. Yet this was not the case in the seventeenth century. As a matter of fact both interpretations fit the optical data equally well and a man of that period had no scientific, but only philosophical, reasons for choosing between them. And it was for philosophical reasons that the prevailing trend of seventeenth century thinking rejected the first interpretation and adopted the second.

There is no doubt whatever that Galileo adopted it too, though he very seldom asserts it. As a matter of fact he does it only once, in a curious passage of his *Letter to Ingoli* where he tells the latter that: [7]

> If it is true, as is commonly held,[8] that the highest parts of the universe are reserved for the habitation of substances more pure and perfect [than ourselves] they [the fixed stars] will be no less lucid and resplendent than the sun; and yet their light, and I mean the light of all of them taken together, does not come up to the tenth part of the visible magnitude and of the light that is communicated by the sun; and of the one as well as of the other of these effects the sole reason is their great distance: how great therefore must we not believe it to be?

Indeed, in the debate about the finiteness or the infinity of the universe, the great Florentine, to whom modern science owes perhaps more than to any other man, takes no part. He never tells us whether he believes the one or the other. He seems not to have made up his mind, or even, though inclining towards infinity, to consider the question as being insoluble. He does not hide, of course, that in contradistinction to Ptolemy, Copernicus and Kepler, he does not admit the limitation of the world or its enclosure by a real *sphere* of fixed stars. Thus in the letter to Ingoli already quoted he tells him: [9]

> You suppose that the stars of the firmament are, all of them, placed in the same orb: that is something the knowledge of which is so doubtful that it will never be proved either by you or by anybody else; but if we restrict ourselves to conjectures and probabilities I shall say that not even four of the fixed stars . . . are at the same distance

> from whichever point of the universe you may want to choose.

And, what is more, not only is it not proved that they are arranged in a sphere but neither Ingoli himself,[10]

> ... nor any one in the world, knows, nor can possibly know, not only what is the shape [of the firmament] but even whether it has any figure at all.

Consequently, once more in opposition to Ptolemy, Copernicus and Kepler, and in accordance with Nicholas of Cusa and Giordano Bruno, Galileo rejects the conception of a center of the universe where the earth, or the sun, should be placed, " the center of the universe which we do not know where to find or whether it exists at all." He even tells us that " the fixed stars are so many suns." Yet, in the selfsame *Dialogue on the Two Greatest World-Systems* from which the last two quotations are taken, discussing *ex professo* the distribution of the fixed stars in the universe, he does not assert that the stars are scattered in space without end: [11]

> SALV. — Now, Simplicius, what shall we do with the fixed stars? Shall we suppose them scattered through the immense abysses of the universe, at different distances from one determinate point; or else placed in a surface spherically distended about a center of its own, so that each of them may be equidistant from the said center?
>
> SIMP. — I would rather take a middle way and would assign them a circle described about a determinate center and comprised within two spherical surfaces, to wit, one very high and concave, the other lower and convex betwixt which I would constitute the innumerable multitude of stars, but yet at diverse altitudes, and this might be called

the sphere of the universe, containing within it the circles of the planets already by us described.

SALV. — But now we have all this while, Simplicius, disposed the mundane bodies exactly according to the order of Copernicus. . . .

We can assuredly explain the moderation of Salviati, who does not criticize the conception presented by Simplicio — though he does not share it — and who accepts it, for the purpose of the discussion, as agreeing perfectly with Copernican astronomy, by the very nature of the *Dialogue*: a book intended for the "general reader," a book which aims at the destruction of the Aristotelian world-view in favor of that of Copernicus, a book which pretends, moreover, not to do it, and where, therefore, subjects both difficult and dangerous are obviously to be avoided.

We could even go as far as to discard the outright negation of the infinity of space in the *Dialogue* — which had to pass the censorship of the Church — and to oppose to it the passage of the letter to Ingoli where its possibility is just as strongly asserted. In the *Dialogue*, indeed, Galileo tells us, just as Kepler does, that it is: [12]

. . . absolutely impossible that there should be an infinite space superior to the fixed stars, for there is no such place in the world; and if there were, the star there situated would be imperceptible to us.

Whereas in the *Letter to Ingoli* he writes: [13]

Don't you know that it is as yet undecided (and I believe that it will ever be so for human knowledge) whether the universe is finite or, on the contrary, infinite. And, given that it be truly infinite, how would you be able to say that the magnitude of the stellar sphere would be proportionate to that of the *orbis magnum*, if this one, in respect to the universe, were rather smaller than a grain of millet in respect to it?

We must not forget, however, that in the selfsame *Dialogue* where he so energetically denied the infinity of space, he makes Salviati tell Simplicio — just as he himself had told Ingoli — that: [14]

> Neither you nor any one else has ever proved that the world is finite and figurate or else infinite and interminate.

Moreover, we cannot reject the testimony of Galileo's *Letter to Liceti*, where, coming back to the problem of the finiteness and the infinity of the world, he writes: [15]

> Many and subtle reasons are given for each of these views but none of them, to my mind, leads to a necessary conclusion, so that I remain in doubt about which of the two answers is the true one. There is only one particular argument of mine that inclines me more to the infinite and interminate than to the terminate (note that my imagination is of no help here since I cannot imagine it either finite or infinite): I feel that my incapacity to comprehend might more properly be referred to incomprehensible infinity, rather than to finiteness, in which no principle of incomprehensibility is required. But this is one of those questions happily inexplicable to human reason, and similar perchance to predestination, free-will and such others in which only Holy Writ and divine revelation can give an answer to our reverent remarks.

It is possible, of course, that *all* the pronouncements of Galileo have to be taken *cum grano salis*, and that the fate of Bruno, the condemnation of Copernicus in 1616, his own condemnation in 1633 incited him to practise the virtue of prudence: he never mentions Bruno, either in his writings or in his letters; yet it is also possible — it is even quite probable — that this problem, like, generally speaking, the problems of cosmology or even of celestial mechanics, did not interest him very much. Indeed he concentrates on the question: *a quo moventur projecta?* but never asks: *a quo moventur planetae?* It may be, therefore, that, like Copernicus himself, he never took up the question, and thus never made the decision — though it is implied in the geometrization of space of which he was one of the foremost promoters — to make his world infinite. Some features of his dynamics, the fact that he never could completely free himself from the obsession of circularity — his planets move circularly around the sun without developing any centrifugal force in their motion — seem to suggest that his world was not infinite. If it was not finite it was probably, like the world of Nicholas of Cusa, indeterminate; and it is, perhaps, more than a pure contingent coincidence that in his letter to Liceti he uses the expression also employed by Cusa: *interminate.*

Be this as it may, it is not Galileo, in any case, nor Bruno, but Descartes who clearly and distinctly formulated principles of the new science, its dream *de reductione scientiae ad mathematicam,* and of the new, mathematical, cosmology. Though, as we shall see, he overshot the mark and by his premature identification of matter and space deprived himself of the means of giving a correct solution to the problems that seventeenth century science had placed before him.

The God of a philosopher and his world are correlated. Now Descartes' God, in contradistinction to most previous Gods, is not symbolized by the things He created; He does not express Himself in them. There is no analogy between God and the world; no *imagines* and *vestigia Dei in mundo*; the only exception is our soul, that is, a pure mind, a being, a substance of which all essence consists in thought, a mind endowed with an intelligence able to grasp the idea of God, that is, of the infinite (which is even innate to it), and with will, that is, with infinite freedom. The Cartesian God gives us some clear and distinct ideas that enable us to find out the truth, provided we stick to them and take care not to fall into error. The Cartesian God is a truthful God; thus the knowledge about the world created by Him that our clear and distinct ideas enable us to reach is a true and authentic knowledge. As for this world, He created it by pure will, and even if He had some reasons for doing it, these reasons are only known to Himself; we have not, and cannot have, the slightest idea of them. It is therefore not only hopeless, but even preposterous to try to find out His aims. Teleological conceptions and explanations have no place and no value in physical science, just as they have no place and no meaning in mathematics, all the more so as the world created by the Cartesian God, that is, the world of Descartes, is by no means the colorful, multiform and qualitatively determined world of the Aristotelian, the world of our daily life and experience — that world is only a subjective world of unstable and inconsistent opinion based upon the untruthful testimony of confused and erroneous sense-perception — but a strictly uniform mathematical world, a world of geometry made real about which our clear and distinct ideas give us a certain and evident knowledge. There is nothing else in this world but matter and motion; or, matter being identical with space or extension, there is nothing else but extension and motion.

The famous Cartesian identification of extension and matter (that is, the assertion that "it is not heaviness, or hardness, or color which constitutes the nature of body but only extension,"[16] in other words, that "nature of body, taken generally, does not consist in the fact that it is a hard, or a heavy, or a colored thing, or a thing that touches our senses in any other manner, but only in that it is a *substance* extended in length, breadth and depth," and that conversely, extension in length, breadth and depth can only be conceived — and therefore can only exist — as belonging to a *material substance*) implies very far-reaching consequences, the first being the negation of the void, which is rejected by Descartes in a manner even more radical than by Aristotle himself.

Indeed, the void, according to Descartes, is not only *physically impossible*, it is essentially impossible. Void space — if there were anything of that kind — would be a *contradictio in adjecto*, an existing nothing. Those who assert its existence, Democritus, Lucretius and their followers, are victims of false imagination and confused thinking. They do not realize that *nothing* can have no properties and therefore no dimensions. To speak of ten feet of void space separating two bodies is meaningless: if there were a void, there would be no separation, and bodies separated *by nothing* would be in contact. And if there is separation and distance, this distance is not a length, breadth or depth of *nothing* but of something, that is, of substance or matter, a "subtle" matter, a matter that we do not sense — that is precisely why people who are accustomed to imagining instead of thinking speak of *void* space — but nevertheless a matter just as real and as "material" (there are no degrees in materiality) as the "gross" matter of which trees and stones are made.

Thus Descartes does not content himself with stating, as did Giordano Bruno and Kepler, that there is no really void space in the world and that the world-space is everywhere filled with "ether." He goes much farther and denies that there is such a thing at all as "space," an entity distinct from "matter" that "fills" it. Matter and space are identical and can be distinguished only by abstraction. Bodies are not *in space,* but only among other bodies; the space that they "occupy" is not anything different from themselves: [17]

> The space or the interior locus, and the body which is comprised in this space are not distinct except in our thought. For, as a matter of fact, the same extension in length, breadth and depth that constitutes space, constitutes also body; and the difference between them consists only in this, that we attribute to body a particular extension, which we conceive to change place with it every time that it is transported, and that we attribute to space an [extension] so general and so vague, that after having removed from a certain space the body which occupied it, we do not think that we have also transported the extension of that space, because it seems to us that the same extension remains there all the time, as long as it is of the same magnitude, of the same figure and has not changed its situation in respect to the external bodies by means of which we determine it.

But that, of course, is an error. And,[18]

> . . . it will be easy to recognize that the same extension that constitutes the nature of body constitutes also the nature of space so that they do not differ in any other way than the nature of the gender or of the species differs from the nature of the individual.

We can, indeed, divest and deprive any given body of all its sensible qualities and [19]

> . . . we shall find that the true idea we have of it consists in this alone, that we perceive distinctly that it is a substance extended in length, breadth and depth. But just that is comprised in the idea we have of space, not only of that which is full of bodies, but also that one which is called void.

Thus,[20]

> . . . the words " place " and " space " do not signify anything which differs *really* from the body that we say to be in some place, and denote only its magnitude, its figure and the manner in which it is situated among other bodies.

Consequently,[21]

> . . . there cannot be any void in the sense in which philosophers take this word, namely as denoting a space where there is no substance, and it is evident that there is no space in the universe that would be such, because the extension of space or of the interior locus is not different from the extension of the body. And as from this alone, that a body is extended in length, breadth and depth, we have reason to conclude that it is a substance, because we conceive that it is not possible that that which is nothing should have an extension, we must conclude the same about the space

supposed to be void: namely that, as there is in it some extension, there is necessarily also some substance.

The second important consequence of the identification of extension and matter consists in the rejection not only of the finiteness and limitation of space, but also that of the real material world. To assign boundaries to it becomes not only false, or even absurd, but contradictory. We cannot posit a limit without transcending it in this very act. We have to acknowledge therefore that the real world is infinite, or rather — Descartes, indeed, refuses to use this term in connection with the world — *indefinite*.

It is clear, of course, that we cannot limit Euclidean space. Thus Descartes is perfectly right in pursuing: [22]

> We recognize moreover that this world, or the entirety of the corporeal substance, has no limits in its extension. Indeed, wherever we imagine such limits, we always not only imagine beyond them some indefinitely extended spaces, but we even perceive them to be truly imaginable, that is, real; and therefore to contain in them also the indefinitely extended corporeal substance. This because, as we have already sufficiently shown, the idea of this extension which we conceive in such a space is obviously identical with that of the corporeal substance itself.

There is no longer any need to discuss the question whether fixed stars are big or small, far or near; more exactly this problem becomes a factual one, a problem of astronomy and observational technics and calculation. The question no longer has metaphysical meaning since it is perfectly certain that, be the stars far or near, they are, like ourselves and our sun, in the midst of other stars without end.

It is exactly the same concerning the problem of the constitution of the stars. This, too, becomes a purely scientific, factual question. The old opposition of the earthly world of change and decay to the changeless world of the skies which, as we have seen, was not abolished by the Copernican revolution, but persisted as the opposition of the moving world of the sun and the planets to the motionless, fixed stars, disappears without trace. The unification and the uniformization of the universe in its contents and laws becomes a self-evident fact [23] — " The matter of the sky and of the earth is one and the same; and there cannot be a plurality of worlds " — at least if we take the term " world " in its full sense, in which it was used by Greek and mediaeval tradition, as meaning a complete and self-centered whole. The world is not an unconnected multiplicity of such wholes utterly separated from each other: it is a unity in which — just as in the universe of Giordano Bruno (it is a pity that Descartes does not use Bruno's terminology) — there are an infinite number of subordinate and interconnected systems, such as our system with its sun and planets, immense vortices of matter everywhere identical joining and limiting each other in boundless space.[24]

> It is easy to deduce that the matter of the sky is not different from that of the earth; and generally, even if the worlds were infinite, it is impossible that they should not be constituted from one and the same matter; and therefore, they cannot be many, but only one: because we understand clearly that this matter of which the whole of nature consists, being an extended substance, must already occupy completely all the imaginary spaces in which these other

worlds should be; and we do not find in ourselves the idea of any other matter.

The infinity of the world seems thus to be established beyond doubt and beyond dispute. Yet, as a matter of fact, Descartes never asserts it. Like Nicholas of Cusa two centuries before him, he applies the term "infinite" to God alone. God is infinite. The world is only *indefinite.*

The idea of the infinite plays an important part in the philosophy of Descartes, so important that Cartesianism may be considered as being wholly based upon that idea. Indeed, it is only as an absolutely infinite being that God can be conceived; it is only as such that He can be proved to exist; it is only by the possession of this idea that man's very nature — that of a finite being endowed with the idea of God — can be defined.

Moreover, it is a very peculiar, and even unique, idea: it is certainly a *clear* and *positive* one — we do not reach infinity by negating finitude; on the contrary, it is by negating the infinite that we conceive finiteness, and yet it is not *distinct.* It so far surpasses the level of our finite understanding that we can neither comprehend nor even analyse it completely. Descartes thus rejects as perfectly worthless all the discussions about the infinite, especially those *de compositione continui,* so popular in the late Middle Ages, and also in the xviith century. He tells us that: [25]

We must never dispute about the infinite, but only hold those things to which we do not find any limit, such as the extension of the world, the divisibility of the parts of matter, the number of stars, etc., to be indefinite.

> Thus we shall never burden ourselves with disputes about the infinite. Indeed, as we are finite, it would be absurd for us to want to determine anything about it, to comprehend it, and thus to attempt to make it *quasi*-finite. Therefore we shall not bother to answer those who would inquire whether, if there were an infinite line, its half would also be infinite; or whether an infinite number would be even or odd; and such like; because about them nobody seems to be able to think except those who believe that their mind is infinite. As for us, in regard to those [things] to which in some respects we are not able to assign any limit, we shall not assert that they are infinite, but we shall consider them as indefinite. Thus, because we cannot imagine an extension so great that a still greater one could not be conceived, we shall say that the magnitude of possible things is indefinite. And because a body cannot be divided into so many parts that further division would not be conceivable, we shall admit that quantity is indefinitely divisible. And because it is impossible to imagine such a number of stars that we should believe God could not create still more, we shall assume that their number is indefinite.

In this way we shall avoid the Keplerian objections based upon the absurdity of an actually infinite distance between ourselves and a given star, and also the theological objections against the possibility of an actually infinite creature. We shall restrict ourselves to the assertion that, just as in the series of numbers, so in world-extension we can always go on without ever coming to an end: [26]

> All these [things] we shall call indefinite rather than infinite: on the one hand that we may reserve the concept of infinity for God alone, because in Him alone we not only do not recognize any limits whatsoever, but also understand

> positively that there are none; and on the other hand because, concerning these things, we do not understand in the same positive way that, in certain respects, they have no limits, but only in a negative way that their limits, if they had any, cannot be found by us.

The Cartesian distinction between the infinite and the indefinite thus seems to correspond to the traditional one between actual and potential infinity, and Descartes' world, therefore, seems to be only potentially infinite. And yet . . . what is the exact meaning of the assertion that the limits of the world cannot be found by us? Why can they not? Is it not, in spite of the fact that we do not understand it in a positive way, simply because there are none? Descartes, it is true, tells us that God alone is clearly understood by us to be infinite and infinitely, that is absolutely, perfect. As for other things: [27]

> We do not recognize them to be so absolutely perfect, because, though we sometimes observe in them properties that seem to us to have no limits, we do not fail to recognize that this proceeds from the defect of our understanding and not from their nature.

But it is hard to admit that the impossibility of conceiving a limit to space must be explained as a result of a defect of our understanding, and not as that of an insight into the nature of the extended substance itself. It is even harder to believe that Descartes himself could seriously espouse this opinion, that is, that *he* could really think that *his* inability to conceive, or even imagine, a finite world could be explained in this way. This is all the more so as somewhat farther on, in the beginning of the third part of the *Principia Philosophiae*, from which the passages we have quoted are taken, we find Descartes telling us that in order to avoid error,[28]

> We have to observe two things carefully: the first being that we always keep before our eyes that God's power and goodness are infinite, in order that this should make us understand that we must not fear to fail in imagining His works too great, too beautiful or too perfect; but that, on the contrary, we can fail if we suppose in them any boundaries or limits of which we have certain knowledge.

The second of these necessary precautions is that,[29]

> We must always keep before our eyes that the capacity of our mind is very mediocre, and that we must not be so presumptuous as it seems we should be if we supposed that the universe had any limits, without being assured of it by divine revelation or, at least, by very evident natural reasons; because it would [mean] that we want our thoughts to be able to imagine something beyond that to which God's power has extended itself in creating the world. . . .

which seems to teach us that the limitations of our reason manifest themselves in assigning limits to the world, and not in denying outright their existence. Thus, in spite of the fact that Descartes, as we shall see in a moment, had really very good reasons for opposing the " infinity " of God to the " indefiniteness " of the world, the common opinion of his time held that it was a pseudo-distinction, made for the purpose of placating the theologians.

That is, more or less, what Henry More, the famous Cambridge Platonist and friend of Newton, was to tell him.

V. Indefinite Extension or Infinite Space

Descartes

& Henry More

Henry More was one of the first partisans of Descartes in England even though, as a matter of fact, he never was a Cartesian and later in life turned against Descartes and even accused the Cartesians of being promoters of atheism.[1] More exchanged with the French philosopher a series of extremely interesting letters which throws a vivid light on the respective positions of the two thinkers.[2]

More starts, naturally, by expressing his admiration for the great man who has done so much to establish truth and dissipate error, continues by complaining about the difficulty he has in understanding some of his teachings, and ends by presenting some doubts, and even some objections.

Thus, it seems to him difficult to understand or to admit the radical opposition established by Descartes between body and soul. How indeed can a purely spiritual soul, that is, something which, according to Descartes, has no extension whatever, be joined to a purely material body, that is, to something which is only and solely extension? Is it not better to assume that the soul, though immaterial, is also extended; that everything, even God, is extended? How could He otherwise be present in the world?

Thus More writes: [3]

> *First*, you establish a definition of matter, or of body, which is much too wide. It seems, indeed, that God is an extended thing (*res*), as well as the Angel; and in general everything that subsists by itself, so that it appears that extension is enclosed by the same limits as the absolute essence of things, which however can vary according to the variety of these very essences. As for myself, I believe it to be clear that God is extended in His manner just because He is omnipresent and occupies intimately the whole machine of the world as well as its singular particles. How indeed could He communicate motion to matter, which He did once, and which, according to you, He does even now, if He did not touch the matter of the universe in practically the closest manner, or at least had not touched it at a certain time? Which certainly He would never be able to do if He were not present everywhere and did not occupy all the spaces. God, therefore, extends and expands in this manner; and is, therefore, an extended thing (*res*).

Having thus established that the concept of extension cannot be used for the definition of matter since it is too wide and embraces *both* body and spirit which *both* are extended, though in a different manner (the Cartesian demonstration of the contrary appears to More to be not only false but even pure sophistry), More suggests *secondly* that matter, being necessarily sensible, should be defined only by its relation to sense, that is, by tangibility. But if Descartes insists on avoiding all reference to sense-perception, then matter should be defined by the ability of bodies to be in mutual contact, and by the impenetrability which matter possesses in contradistinction to spirit. The latter, though extended, is freely penetrable and cannot be touched. Thus spirit and body can co-exist in the same place, and, of course, two — or any number of — spirits can have the same identical location and " penetrate " each other, whereas for bodies this is impossible.

The rejection of the Cartesian identification of extension and matter leads naturally to the rejection by Henry More of Descartes' denial of the possibility of vacuum. Why should not God be able to destroy all matter contained in a certain vessel without — as Descartes asserts — its walls being obliged to come together? Descartes, indeed, explains that to be separated by " nothing " is contradictory and that to attribute dimensions to " void " space is exactly the same as to attribute properties to nothing; yet More is not convinced, all the more so as " learned Antiquity " — that is Democritus, Epicurus, Lucretius — was of quite a different opinion. It is possible, of course, that the walls of the vessel will be brought together by the pressure of matter outside them. But if that happens, it will be because of a natural necessity and not because of a logical one. Moreover, this void space will not be absolutely void, for it will continue to be filled with God's extension. It will only be void of matter, or body, properly speaking.

In the *third* place Henry More does not understand the " singular subtlety " of Descartes' negation of the existence of atoms, of his assertion of the indefinite divisibility of matter, combined with the use of corpuscular conceptions in his own physics. To say that the admission of atoms is limiting God's omnipotence, and that we cannot deny that God could, if He wanted to, divide the atoms into parts, is of no avail: the indivisibility of atoms means their indivisibility by any created power, and that is something that is perfectly compatible with God's own power to divide them, *if* He wanted to do so. There are a great many things that He could have done, but did not, or even those that He can do but does not. Indeed, if God wanted to preserve his omnipotence in its absolute status, He would never create matter at all: for, as matter is always divisible into parts that are themselves divisible, it is clear that God will never be able to bring this division to its end and that there will always be something which evades His omnipotence.

Henry More is obviously right and Descartes himself, though insisting on God's omnipotence and refusing to have it limited and bounded even by the rules of logic and mathematics, cannot avoid declaring that there are a great many things that God cannot do, either because to do them would be, or imply, an imperfection (thus, for instance, God cannot lie and deceive), or because it would make no sense. It is just because of that, Descartes asserts, that even God cannot make a void, or an atom. True, according to Descartes, God could have created quite a different world and could have made twice two equal to five, and not to four. On the other hand, it is equally true that He did not do it and that *in this world* even God cannot make twice two equal to anything but four.

From the general trend of his objections it is clear that the Platonist, or rather Neoplatonist, More was deeply influenced by the tradition of Greek atomism, which is not surprising in view of the fact that one of his earliest works bears the revealing title, *Democritus Platonissans.* . . .[4]

What he wants is just to avoid the Cartesian geometrization of being, and to maintain the old distinction between *space* and the things that are *in space*; that are moving *in space* and not only relatively to each other; that *occupy* space in virtue of a special and proper quality or force — impenetrability — by which they resist each other and exclude each other from their " places."

Grosso modo, these are Democritian conceptions and that explains the far-reaching similarity of Henry More's objections to Descartes to those of Gassendi, the chief representative of atomism in the XVIIth century.[5] Yet Henry More is by no means a pure Democritian. He does not reduce being to matter. And his space is not the infinite void of Lucretius: it is full, and not full of " ether " like the infinite space of Bruno. It is full of God, and in a certain sense it is God Himself as we shall see more clearly hereafter.

Let us now come to More's *fourth* and most important objection to Descartes: [6]

> *Fourth,* I do not understand your indefinite extension of the world. Indeed this indefinite extension is either *simpliciter* infinite, or only in respect to us. If you understand extension to be infinite *simpliciter*, why do you obscure your thought by too low and too modest words? If it is infinite only in respect to us, extension, in reality, will be finite; for our mind is the measure neither of the things nor of truth. And therefore, as there is another *simpliciter* infinite expansion, that of the divine essence, the matter of your

> vortices will recede from their centers and the whole fabric of the world will be dissipated into atoms and grains of dust.[7]

Having thus impaled Descartes on the horns of the dilemma, More continues: [8]

> I admire all the more your modesty and your fear of admitting the infinity of matter as you recognize, on the other hand, that matter is divided into an actually infinite number of particles. And if you did not, you could be compelled to do so,

by arguments that Descartes would be bound to accept.[9]

To the perplexity and objections of his English admirer and critic Descartes replies [10] — and his answer is surprisingly mild and courteous — that it is an error to define matter by its relation to senses, because by doing so we are in danger of missing its true essence, which does not depend on the existence of men and which would be the same if there were no men in the world; that, moreover, if divided into sufficiently small parts, all matter becomes utterly insensible; that his proof of the identity of extension and matter is by no means a sophism but is as clear and demonstrative as it could be; and that it is perfectly unnecessary to postulate a special property of impenetrability in order to define matter because it is a mere consequence of its extension.

Turning then to More's concept of immaterial or spiritual extension, Descartes writes: [11]

> I am not in the habit of disputing about words, and therefore if somebody wants to say that God is, in some sense, extended because He is everywhere, I shall not

> object. But I deny that there is in God, in an Angel, in our soul, and in any substance that is not a body, a true extension, such as is usually conceived by everybody. For by an extended thing everybody understands something [which is] imaginable (be it an *ens rationis* or a real thing), and in which, by imagination, can be distinguished different parts of a determined magnitude and figure, of which the one is in no way the other; so that it is possible, by imagination, to transfer any one of them to the place of another, but not to imagine two of them in the same place.

Nothing of that kind applies to God, or to our souls, which are not objects of imagination, but of pure understanding, and have no separable parts, especially no parts of determinate size and figure. Lack of extension is precisely the reason why God, the human soul, and any number of angels can be all together in the same place. As for atoms and void, it is certain that, our intelligence being finite and God's power infinite, it is not proper for us to impose limits upon it. Thus we must boldly assert "that God can do all that we conceive to be possible, but not that He cannot do what is repugnant to our concept." Nevertheless, we can judge only according to our concepts, and, as it is repugnant to our manner of thinking to conceive that, if all matter were removed from a vessel, extension, distance, etc., would still remain, or that parts of matter be indivisible, we say simply that all that implies contradiction.

Descartes' attempt to save God's omnipotence and, nevertheless, to deny the possibility of void space as incompatible with our manner of thinking, is, to say the truth, by no means convincing. The Cartesian God is a *Deus verax* and He guarantees the truth of our clear and distinct ideas. Thus it is not only repugnant to our thought, but impossible that something of which we clearly see that it implies contradiction be real. There are no contradictory objects in this world, though there could have been in another.

Coming now to More's criticism of his distinction between " infinite " and " indefinite," Descartes assures him that it is not because of [12]

> . . . an affectation of modesty, but as a precaution, and, in my opinion a necessary one, that I call certain things indefinite rather than infinite. For it is God alone whom I understand positively to be infinite; as for the others, such as the extension of the world, the number of parts into which matter is divisible, and so on, whether they are *simpliciter* infinite or not, I confess not to know. I only know that I do not discern in them any end, and therefore, in respect to me, I say they are indefinite. And though our mind is not the measure of things or of truth, it must, assuredly, be the measure of things that we affirm or deny. What indeed is more absurd or more inconsiderate than to wish to make a judgment about things which we confess to be unable to perceive with our mind?
>
> Thus I am surprised that you not only seem to want to do so, as when you say that *if* extension is *infinite only in respect to us then extension in truth will be finite*, etc., but that you imagine beyond this one a certain divine extension, which would stretch farther than the extension of bodies, and thus suppose that God has *partes extra partes*, and that He is divisible, and, in short, attribute to Him all the essence of a corporeal being.

Descartes, indeed, is perfectly justified in pointing out that More has somewhat misunderstood him: a space *beyond* the world of extension has never been admitted by him as possible or imaginable, and even if the world *had* these limits which we are unable to find, there certainly would be nothing beyond them, or, better to say, there would be no *beyond*. Thus, in order to dispel completely More's doubts, he declares: [13]

> When I say that the extension of matter is indefinite, I believe it to be sufficient to prevent any one imagining a place outside it, into which the small particles of my vortices could escape; because wherever this place be conceived, it would already, in my opinion, contain some matter; for, when I say that it is indefinitely extended, I am saying that it extends farther than all that can be conceived by man.
>
> But I think, nevertheless, that there is a very great difference between the amplitude of this corporeal extension and the amplitude of the divine, I shall not say, extension, because properly speaking there is none, but substance or essence; and therefore I call this one *simpliciter* infinite, and the other, indefinite.

Descartes is certainly right in wanting to maintain the distinction between the "intensive" infinity of God, which not only excludes all limit, but also precludes all multiplicity, division and number, from the mere endlessness, indefiniteness, of space, or of the series of numbers, which necessarily include and presuppose them. This distinction, moreover, is quite traditional, and we have seen it asserted not only by Nicholas of Cusa, but even by Bruno.

Henry More does not deny this distinction; at least not completely. In his own conception it expresses itself in the opposition between the material and the divine extension. Yet, as he states it in his second letter to

Descartes,[14] it has nothing to do with Descartes' assertion that there may be limits to space and with his attempt to build a concept intermediate between the finite and the infinite; the world is finite or infinite, *tertium non datur*. And if we admit, as we must, that God is infinite and everywhere present, this "everywhere" can only mean infinite space. In this case, pursues More, re-editing an argument already used by Bruno, there must also be matter everywhere, that is, the world must be infinite.[15]

> You can hardly ignore that it is either *simpliciter* infinite or, in point of fact, finite, though you cannot as easily decide whether it is the one or the other. That, however, your vortices are not disrupted and do not come apart seems to be a rather clear sign that the world is really infinite. For my part, I confess freely that though I can boldly give my approval to this axiom: *The world is finite, or not finite*, or, what is here the same thing, *infinite*, I cannot, nevertheless, fully understand the infinity of any thing whatsoever. But here there comes to my imagination what Julius Scaliger wrote somewhere about the contraction and the dilatation of the Angels: namely, that they cannot extend themselves *in infinitum*, or contract themselves to an imperceptible (*οὐδενότητα*) point. Yet if one recognizes God to be positively infinite (that is, existing everywhere), as you yourself rightly do, I do not see whether it is permitted to the unbiassed reason to hesitate to admit forthwith also that He is nowhere idle, and that with the same right, and with the same facility with which [He created] this matter in which we live, or that to which our eyes and our mind can reach, He produced matter everywhere.

Nor is it absurd or inconsiderate to say that, if the extension is infinite only *quoad nos*, it will, in truth and in reality, be finite: [16]

> I will add that this consequence is perfectly manifest, because the particle " only " (*tantum*) clearly excludes all real infinity of the thing which is said to be infinite only in respect to us, and therefore in reality the extension will be finite; moreover my mind does perceive these things of which I judge, as it is perfectly clear to me that the world is either finite or infinite, as I have just mentioned.

As for Descartes' contention that the impossibility of the void already results from the fact that " nothing " can have no properties or dimensions and therefore cannot be measured, More replies by denying this very premise: [17]

> . . . for, if God annihilated this universe and then, after a certain time, created from nothing another one, this *intermundium* or this absence of the world would have its duration which would be measured by a certain number of days, years or centuries. There is thus a duration of something that does not exist, which duration is a kind of extension. Consequently, the amplitude of nothing, that is of void, can be measured by ells or leagues, just as the duration of what does not exist can be measured in its inexistence by hours, days and months.

We have seen Henry More defend, against Descartes, the infinity of the world, and even tell the latter that his own physics necessarily implies this infinity. Yet it seems that, at times, he feels himself assailed by doubt. He is perfectly sure that space, that is, God's extension, is infinite. On the other hand, the material world may, perhaps, be finite. After all, nearly everybody believes it; spatial infinity and temporal eternity are strictly parallel, and so both seem to be absurd. Moreover Cartesian cosmology can be put in agreement with a finite world. Could

Descartes not tell what would happen, in this case, if somebody sitting at the extremity of the world pushed his sword through the limiting wall? On the one hand, indeed, this seems easy, as there would be nothing to resist it; on the other, impossible, as there would be no place where it could be pushed.[18]

Descartes' answer to this second letter of More [19] is much shorter, terser, less cordial than to the first one. One feels that Descartes is a bit disappointed in his correspondent who obviously does not understand his, Descartes', great discovery, that of the essential opposition between mind and extension, and who persists in attributing extension to souls, angels, and even to God. He restates [20]

> . . . that he does not conceive any extension of substance in God, in the angels, or in our mind, but only an extension of power, so that an angel can proportionate this power to a greater or smaller part of corporeal substance; for if there were no body at all, this power of God or of an angel would not correspond to any extension whatever. To attribute to substance what pertains only to power is an effect of the same prejudice which makes us suppose all substance, even that of God, to be something that can be imagined.

If there were no world, there would be no time either. To More's contention that the *intermundium* would last a certain time, Descartes replies: [21]

> I believe that it implies a contradiction to conceive a duration between the destruction of the first world and the creation of the second one; for, if we refer this duration or something similar to the succession of God's ideas, this will be an error of our intellect and not a true perception of something.

Indeed, it would mean introducing time into God, and thus making God a temporal, changing being. It would mean denying His eternity, replacing it by mere sempiternity — an error no less grave than the error of making Him an extended thing. For in both cases God is menaced with losing His transcendence, with becoming immanent to the world.

Now Descartes' God is perhaps not the Christian God, but a philosophical one.[22] He is, nevertheless, God, not the soul of the world that penetrates, vivifies and moves it. Therefore he maintains, in accordance with mediaeval tradition, that, in spite of the fact that in God power and essence are one — an identity pointed out by More in favour of God's actual extension — God has nothing in common with the material world. He is a pure mind, an infinite mind, whose very infinity is of a unique and incomparable non-quantitative and non-dimensional kind, of which spatial extension is neither an image nor even a symbol. The world therefore, must not be called infinite; though of course we must not enclose it in limits: [23]

> It is repugnant to my concept to attribute any limit to the world, and I have no other measure than my perception for what I have to assert or to deny. I say, therefore, that the world is indeterminate or indefinite, because I do not recognize in it any limits. But I dare not call it infinite as I perceive that God is greater than the world, not in respect to His extension, because, as I have already said, I do not acknowledge in God any proper [extension], but in respect to His perfection.

Once more Descartes asserts that God's presence in the world does not imply His extension. As for the world itself which More wants to be either *simpliciter* finite, or *simpliciter* infinite, Descartes still refuses to call it infinite. And yet, either because he is somewhat angry with More, or because he is in a hurry and therefore less careful, he practically abandons his former assertion about the possibility of the world's having limits (though we cannot find them) and treats this conception in the same manner in which he treated that of the void, that is, as nonsensical and even contradictory; thus, rejecting as meaningless the question about the possibility of pushing a sword through the boundary of the world, he says: [24]

> It is repugnant to my mind, or what amounts to the same thing, it implies a contradiction, that the world be finite or limited, because I cannot but conceive a space outside the boundaries of the world wherever I presuppose them. But, for me, this space is a true body. I do not care if it is called by others imaginary, and that therefore the world is believed to be finite; indeed, I know from what prejudices this error takes its origin.

Henry More, needless to say, was not convinced — one philosopher seldom convinces another. He persisted, therefore, in believing " with all the ancient Platonists " that all substance, souls, angels and God are extended, and that the world, in the most literal sense of this word, is in God just as God is in the world. More accordingly sent Descartes a third letter,[25] which he answered,[26] and a fourth,[27] which he did not.[28] I shall not attempt to examine them here as they bear chiefly on questions which, though interesting in themselves — for example, the discussion about motion and rest — are outside our subject.

Summing up, we can say that we have seen Descartes, under More's pressure, move somewhat from the position he had taken at first: to assert the indefiniteness of the world, or of space, does not mean, negatively, that perhaps it has limits that we are unable to ascertain; it means, quite positively, that it has none because it would be contradictory to posit them. But he cannot go farther. He has to maintain his distinction, as he has to maintain the identification of extension and matter, if he is to maintain his contention that the physical world is an object of pure intellection and, at the same time, of imagination — the precondition of Cartesian science — and that the world, in spite of its lack of limits, refers us to God as its creator and cause.

Infinity, indeed, has always been the essential character, or attribute, of God; especially since Duns Scotus, who could accept the famous Anselmian *a priori* proof of the existence of God (a proof revived by Descartes) only after he had "colored" it by substituting the concept of the infinite being (*ens infinitum*) for the Anselmian concept of a being than which we cannot think of a greater (*ens quo maius cogitari nequit*). Infinity thus — and it is particularly true of Descartes whose God exists in virtue of the infinite "superabundance of His essence" which enables Him to be His own cause (*causa sui*) and to give Himself His own existence [29] — means or implies being, even necessary being. Therefore it cannot be attributed to creature. The distinction, or opposition, between God and creature is parallel and exactly equivalent to that of infinite and of finite being.

VI. God and Space, Spirit and Matter

Henry More

The breaking off of the correspondence with — and the death of — Descartes did not put an end to Henry More's preoccupation with the teaching of the great French philosopher. We could even say that all his subsequent development was, to a very great extent, determined by his attitude towards Descartes: an attitude consisting in a partial acceptance of Cartesian mechanism joined to a rejection of the radical dualism between spirit and matter which, for Descartes, constituted its metaphysical background and basis.

Henry More enjoys a rather bad reputation among historians of philosophy, which is not surprising. In some sense he belongs much more to the history of the hermetic, or occultist, tradition than to that of philosophy proper; in some sense he is not of his time: he is a spiritual contemporary of Marsilio Ficino, lost in the disenchanted world of the "new philosophy" and fighting a losing battle against it. And yet, in spite of his partially anachronistic standpoint, in spite of his invincible trend towards syncretism which makes him jumble together

Plato and Aristotle, Democritus and the Cabala, the thrice great Hermes and the Stoa, it was Henry More who gave to the new science — and the new world view — some of the most important elements of the metaphysical framework which ensured its development: this because, in spite of his unbridled phantasy, which enabled him to describe at length God's paradise and the life and various occupations of the blessed souls and spirits in their post-terrestrial existence, in spite of his amazing credulity (equalled only by that of his pupil and friend, fellow of the Royal Society, Joseph Glanvill,[1] the celebrated author of the *Scepsis scientifica*), which made him believe in magic, in witches, in apparitions, in ghosts, Henry More succeeded in grasping the fundamental principle of the new ontology, the infinitization of space, which he asserted with an unflinching and fearless energy.

It is possible, and even probable, that, at the time of his *Letters* to Descartes (1648), Henry More did not yet recognize where the development of his conceptions was ultimately to lead him, all the more so as these conceptions are by no means "clear" and "distinct." Ten years later, in his *Antidote against Atheism* [2] and his *Immortality of the Soul* [3] he was to give them a much more precise and definite shape; but it was only in his *Enchiridium metaphysicum,*[4] ten years later still, that they were to acquire their final form.

As we have seen, Henry More's criticism of Descartes' identification of space or extension with matter follows two main lines of attack. On the one hand it seems to him to *restrict* the ontological value and importance of extension by reducing it to the role of an essential attribute of matter alone and denying it to spirit, whereas it is an attribute of being as such, the necessary precondition of any real existence. There are not, as Descartes asserts, two types of substance, the extended and the unextended. There is only one type: all substance, spiritual as well as material, is extended.

On the other hand, Descartes, according to More, fails to recognize the specific character both of matter and of space, and therefore misses their essential distinction as well as their fundamental relation. Matter is mobile *in* space and by its impenetrability *occupies* space; space is not mobile and is unaffected by the presence, or absence, of matter in it. Thus matter without space is unthinkable, whereas space without matter, Descartes notwithstanding, is not only an easy, but even a necessary idea of our mind.

Henry More's pneumatology does not interest us here; still, as the notion of spirit plays an important part in his — and not only his — interpretation of nature, and is used by him — and not only by him — to explain natural processes that cannot be accounted for or " demonstrated " on the basis of purely mechanical laws (such as magnetism, gravity and so on), we shall have to dwell for a moment on his concept of it.

Henry More was well aware that the notion of " spirit " was, as often as not, and even more often than not, presented as impossible to grasp, at least for the human mind,[5]

> But for mine own part, I think the *nature* of a *spirit* is as conceivable and easy to be defined as the nature of anything else. For as for the very *Essence* or bare *Substance* of any thing whatsoever, he is a very Novice in speculation that does not acknowledge that utterly unknowable; but for the *Essential* and *Inseparable Properties*, they are as intelligible and explicable in a Spirit as in any other Subject

whatever. As for example, I conceive the intire *Idea* of a *Spirit* in generall, or at least of all finite, created and subordinate *Spirits*, to consist of these severall powers or properties, viz. *Self-penetration, Self-motion, Self-contraction* and *Dilatation*, and *Indivisibility*; and these are those that I reckon more absolute: I will adde also what has relation to another and that is power of *Penetrating, Moving* and *Altering the Matter*. These *Properties* and *Powers* put together make up the *Notion* and *Idea* of a *Spirit* whereby it is plainly distinguished from a *Body* whose parts cannot penetrate one another, is not *Self-moveable*, nor can *contract* nor *dilate* it self, is *divisible* and *separable* one part from another; but the parts of a *Spirit* can be no more separable, though they be dilated, than you can cut off the *Rayes* of the *Sun* by a pair of Scissors made of pellucid Crystall. And this will serve for the settling of the *Notion* of a *Spirit*. And out of this description it is plain that *Spirit* is a notion of more *Perfection* than a *Body*, and therefore more fit to be an *Attribute* of what is *absolutely Perfect* than a *Body* is.

As we see, the method used by Henry More to arrive at the notion or definition of spirit is rather simple. We have to attribute to it properties opposite or contrary to those of body: penetrability, indivisibility, and the faculty to contract and dilate, that is, to extend itself without loss of continuity, into a smaller or larger space. This last property was for a very long time considered as belonging to matter also, but Henry More, under the conjoint influence of Democritus and Descartes, denies it to matter, or body, which is, as such, incompressible and always occupies the same amount of space.

In *The Immortality of the Soul* Henry More gives us an even clearer account both of his notion of spirit and of the manner in which this notion can be determined. Moreover he attempts to introduce into his definition a sort of terminological precision. Thus, he says,[6] "by *Actual Divisibility* I understand *Discerpibility*, gross tearing or cutting of one part from the other." It is quite clear that this "discerpibility" can only belong to a body and that you cannot tear away and remove a piece of a spirit.

As for the faculty of contraction and dilation, More refers it to the "essential spissitude" of the spirit, a kind of spiritual density, fourth mode, or fourth dimension of spiritual substance that it possesses in addition to the normal three of spatial extension with which bodies are alone endowed.[7] Thus, when a spirit contracts, its "essential spissitude" increases; it decreases, of course, when it dilates. We cannot, indeed, *imagine* the "spissitude" but this "*fourth* Mode," Henry More tells us,[8] "is as easy and familiar to my Understanding as that of the *Three dimensions* to my sense or Phansy."

The definition of spirit is now quite easy: [9]

> I will define therefore a *Spirit* in generall thus: *A substance penetrable and indiscerpible.* The fitness of which definition will be better understood, if we divide *Substance* in generall into these first Kindes, viz. *Body* and *Spirit* and then define Body *A Substance impenetrable and discerpible.* Whence the contrary Kind to this is fitly defined, *A Substance penetrable and indiscerpible.*
>
> Now I appeal to any man that can set aside prejudice, and has the free use of his Faculties, whether every term of the Definition of a *Spirit* be not as intelligible and congruous to Reason, as in that of a *Body*. For the precise Notion of *Substance* is the same in both, in which, I conceive, is com-

prised *Extension* and *Activity* either connate or communicated. For *Matter* it self once moved can move other *Matter*. And it is as easy to understand what *Penetrable* is as *Impenetrable*, and what *Indiscerpible* is as *Discerpible*; and *Penetrability* and *Indiscerpibility* being as *immediate* to *Spirit* as *Impenetrability* or *Discerpibility* to *Body*, there is as much reason to be given for the Attributes of the one as of the other, by Axiome 9.[10] And *Substance* in its precise notion including no more of *Impenetrability* than of *Indiscerpibility* we may as well wonder how one kind of Substance holds out its parts one from another so as to make them *impenetrable* to each other (as *Matter*, for instance does the parts of *Matter*) as that parts of another substance hold so fast together that they are by no means *Discerpible*. And therefore the *holding out* in one being as difficult a business to conceive as the *holding together* in the other, this can be no prejudice to the notion of a Spirit.

I am rather doubtful whether the modern reader — even if he puts aside prejudice and makes free use of his faculties — will accept Henry More's assurance that it is as easy, or as difficult, to form the concept of spirit as that of matter, and whether, though recognizing the difficulty of the latter, he will not agree with some of More's contemporaries in " the confident opinion " that " the very notion of *a Spirit* were a piece of Nonsense and perfect Incongruity." The modern reader will be right, of course, in rejecting More's concept, patterned obviously upon that of a ghost. And yet he will be wrong in assuming it to be pure and sheer nonsense.

In the first place, we must not forget that for a man of the seventeenth century the idea of an extended, though not material, entity was by no means something strange or even uncommon. Quite the contrary: these entities were represented in plenty in their daily life as well as in their scientific experience.

To begin with, there was light, assuredly immaterial and incorporeal but nevertheless not only extending through space but also, as Kepler does not fail to point out, able, in spite of its immateriality, to act upon matter, and also to be acted upon by the latter. Did not light offer a perfect example of penetrability, as well as of penetrating power? Light, indeed, does not hinder the motion of bodies through it, and it can also pass through bodies, at least some of them; furthermore, in the case of a transparent body traversed by light, it shows us clearly that matter and light can coexist in the same place.

The modern development of optics did not destroy but, on the contrary, seemed to confirm this conception: a real image produced by mirrors or lenses has certainly a determinate shape and location in space. Yet, is it body? Can we disrupt or " discerp " it, cut off and take away a piece of this image?

As a matter of fact, light exemplifies nearly all the properties of More's " spirit," those of " condensation " and " dilatation " included, and even that of " essential spissitude " that could be represented by the intensity of light's varying, just like the " spissitude," with its " contraction " and " dilatation."

And if light were not sufficiently representative of this kind of entity, there were magnetic forces that to William Gilbert seemed to belong to the realm of animated much more than to purely material being: [11] there was attraction (gravity) that freely passed through *all* bodies and could be neither arrested nor even affected by any.

Moreover, we must not forget that the "ether," which played such an important role in the physics of the nineteenth century (which maintained as firmly or even more firmly than the seventeenth the opposition between "light" and "matter," an opposition that is by no means completely overcome even now), displayed an ensemble of properties even more astonishing than the "spirit" of Henry More. And finally, that the fundamental entity of contemporary science, the "field," is something that possesses location and extension, penetrability and indiscerpibility. . . . So that, somewhat anachronistically, of course, one could assimilate More's "spirits," at least the lowest, unconscious degrees of them, to some kinds of fields.[11a]

But let us now come back to More. The greater precision achieved by him in the determination of the concept of spirit led necessarily to a stricter discrimination between its extension and the space in which, like everything else, it finds itself, concepts that were somehow merged together into the divine or spiritual extension opposed by More to the material Cartesian one. Space or pure immaterial extension will be distinguished now from the "spirit of nature" that pervades and fills it, that acts upon matter and produces the above-mentioned non-mechanical effects, an entity which on the scale of perfection of spiritual beings occupies the very lowest degree. This spirit of nature is [12]

> *A Substance incorporeal but without sense or animadversion, pervading the whole matter of the Universe, and exercising a plastic power therein, according to the sundry predispositions and occasions of the parts it works upon, raising such Phenomena in the world, by directing the parts of the*

matter, and their motion, as cannot be resolved into mere mechanical power.

Among these phenomena unexplainable by purely mechanical forces, of which Henry More knows, alas, a great number, including sympathetic cures and consonance of strings (More, needless to say, is a rather bad physicist), the most important is gravity. Following Descartes, he no longer considers it an essential property of body, or even, as Galileo still did, an unexplainable but real tendency of matter; but — and he is right — he accepts neither the Cartesian nor the Hobbesian explanation of it. Gravity cannot be explained by pure mechanics and therefore, if there were in the world no other, non-mechanical, forces, unattached bodies on our moving earth would not remain on its surface, but fly away and lose themselves in space. That they do not is a proof of the existence in nature of a "more than mechanical," "spiritual" agency.

More writes accordingly in the preface to *The Immortality of the Soul,*[13]

> *I have not only confuted their* [Descartes' and Hobbes'] *Reasons, but also from* Mechanical *principles granted on all sides and confirmed by Experience, demonstrated that the Descent of a stone or a bullet, or any such like heavy Body is enormously contrary to the Laws of* Mechanicks; *and that according to them they would necessarily, if they lye loose, recede from the Earth and be carried away out of our sight into the farthest parts of the Aire, if some* Power more than Mechanical *did not curb that Motion, and force them downwards towards the Earth. So that it is plain that we have not arbitrarily introduced a Principle but that it is forced upon us by the undeniable evidence of Demonstration.*

As a matter of fact the *Antidote against Atheism* had already pointed out that stones and bullets projected upwards return to earth — which, according to the laws of motion, they should not do; for,[14]

> . . . if we consider more particularly what a strong tug a massive Bullet, suppose of lead or brass must needs give (according to that prime *Mechanicall* law of motion persisting in a straight line) to recede from the superficies of the Earth, the Bullet being in so swift a Motion as would dispatch some fifteen Miles in one Minute of an Hour; it must needs appear that a wonderful Power is required to curb it, regulate it, or remand it back to the Earth, and keep it there, notwithstanding the strong Reluctancy of that first Mechanical law of Matter that would urge it to recede. Whereby is manifested not only the marvellous Power of *Unity* in *Indiscerpibility* in the *Spirit of Nature* but that there is a peremptory and even forcible Execution of an *all-comprehensive and eternal Council* for the *Ordering* and the *Guiding* of the Motion of *Matter* in the Universe to what is the *Best*. And this phenomenon of Gravity is of so *good* and *necessary* consequence, that there could be neither Earth nor Inhabitants without it, in this State that things are.

Indeed, without the action of a non-mechanical principle all matter in the universe would divide and disperse; there would not even be bodies, because there would be nothing to hold together the ultimate particles composing them. And, of course, there would be no trace of that purposeful organization which manifests itself not only in plants, animals and so on, but even in the very arrangement of our solar system. All that is the work of the spirit of nature, which acts as an instrument, itself unconscious, of the divine will.

So much for the spirit of nature that pervades the whole universe and extends itself in its infinite space. But what about this space itself? the space that we cannot conceive if not infinite — that is, necessary — and that we cannot " disimagine " (which is a confirmation of its necessity) from our thought? Being immaterial it is certainly to be considered as spirit. Yet it is a " spirit " of quite a special and unique kind, and More is not quite sure about its exact nature. Though, obviously, he inclines towards a very definite solution, namely towards the identification of space with the divine extension itself, he is somewhat diffident about it. Thus he writes: [15]

> If there were no *Matter* but the Immensity of the Divine Essence only occupying all by its Ubiquity, then the *Reduplication*, as I may so speak, of his indivisible substance, whereby he presents himself intirely everywhere, would be the Subject of that Diffusion and Measurability. . . .

for which the Cartesians require the presence of matter, asserting that material extension alone can be measured, an assertion which leads inevitably to the affirmation of the infinity and the necessary existence of matter. But we do not need matter in order to have measures, and More can pursue: [16]

> And I adde further, that the perpetual observation of this infinite Amplitude and Mensurability, which we cannot disimagine in our Phancie but will necessary be, may be a more rude and obscure notion offered to our mind of that *necessary* and *self-existant* Essence which the *Idea* of God does with greater fulness and distinctness represent to us. For it is plain that not so much as our Imagination is engaged to an appropriation of this *Idea of Space* to corporeal

Matter, in that it does not naturally conceive any impenetrability or tangibility in the Notion thereof; and therefore it may as well belong to a *Spirit* as a *Body*. Whence as I said before, the *Idea* of God being such as it is, it will both justly and necessarily cast this ruder notion of *Space* upon that infinite and eternal spirit which is God.

There is also another way of answering this Objection, which is this; that this Imagination of *Space* is not the imagination of any real thing, but only of the large and immense capacity of the potentiality of the *Matter*, which we can not free our Minds from but must necessarily acknowledge that there is indeed such a possibility of Matter to be measured upward, downward, everyway *in infinitum*, whether this *corporeal Matter* were actually there or no; and that though this potentiality of *Matter* and Space be measurable by furloughs, miles, or the like, that it implies no more real Essence or Being, than when a man recounts so many orders or Kindes of the Possibilities of things, the compute or number of them will infer the reality of their Existence.

But if the Cartesians would urge us further and insist upon the impossibility of measuring the nothingness of void space,[17]

. . . it may be answered, That *Distance* is no real or *Physical* property of a thing but only *notional*; because more or less of it may accrue to a thing when as yet there has been nothing at all done to that to which it does accrue.

And if they urge still further and contend, that . . . distance must be some *real* thing . . . I answer briefly that *Distance* is nothing else but the privation of tactual union and the greater *distance* the greater privation . . .; and that this privation of tactual union is measured by *parts*, as other privations of qualities by *degrees*; and that *parts*

> and *degrees*, and such like notions, are not *real* things themselves any where, but our mode of conceiving them, and therefore we can bestow them upon Non-entities as well as Entities. . . .
>
> But if this will not satisfie, 'tis no detriment to our cause. For if after the removal of *corporeal Matter* out of the world, there will be still *Space* and *distance*, in which this very matter, while it was there, was also conceived to lye, and this *distant Space* cannot but be something, and yet not corporeal, because neither impenetrable nor tangible, it must of necessity be a substance Incorporeal, necessarily and eternally existent of it self: which the clearer *Idea* of a *Being absolutely perfect* will more fully and punctually inform us to be the *Self-subsisting* God.

We have seen that, in 1655 and also in 1662, Henry More was hesitating between various solutions of the problem of space. Ten years later his decision is made, and the *Enchiridium metaphysicum* (1672) not only asserts the real existence of infinite void space against all possible opponents, as a real precondition of all possible existence, but even presents it as the best and most evident example of non-material — and therefore spiritual — reality and thus as the first and foremost, though of course not unique, subject-matter of metaphysics.

Thus Henry More tells us that " the first method for proving the uncorporeal things " must be based on [18]

> . . . the demonstration of a certain unmovable extended [being] distinct from the movable matter, which commonly is called *space* or inner *locus*. That it is something real and not imaginary, as many people assert, we shall prove later by various arguments.

Henry More seems to have completely forgotten his own uncertainty concerning the question; in any case he does not mention it and pursues: [19]

> First, it is so obvious that it hardly needs proof, as it is confirmed by the opinions of nearly all the philosophers, and even of all men in general, but particularly of those who, as it is proper, believe that matter was created at a certain time. For we must either acknowledge that there is a certain extended [entity] besides matter, or that God could not create finite matter; indeed, we cannot conceive finite matter but as surrounded on all sides by something infinitely extended.

Descartes remains, as we see, the chief adversary of Henry More; indeed, as More discovered meanwhile, by his denial both of void space and of spiritual extension, Descartes practically excludes spirits, souls, and even God, from his world; he simply leaves no *place* for them in it. To the question "where?," the fundamental question which can be raised concerning any and every real being — souls, spirits, God — and to which Henry More believes he can give definite answers (here, elsewhere or — for God — everywhere), Descartes is obliged, by his principles, to answer: *nowhere, nullibi.* Thus, in spite of his having invented or perfected the magnificent *a priori* proof of the existence of God, which Henry More embraced enthusiastically and was to maintain all his life, Descartes, by his teaching, leads to materialism and, by his exclusion of God from the world, to atheism. From now on, Descartes and the Cartesians are to be relentlessly criticized and to bear the derisive nickname of *nullibists.*

Still, there are not only Cartesians to be combatted.

There is also the last cohort of Aristotelians who believe in a finite world, and deny the existence of space outside it. They, too, have to be dealt with. On their behalf Henry More revives some of the old mediaeval arguments used to demonstrate that Aristotelian cosmology was incompatible with God's omnipotence.

It cannot be doubted, of course, that if the world were finite and limited by a spherical surface with no space outside it,[20]

> it would follow, secondly, that not even divine omnipotence could make it that this corporeal finite world in its ultimate surface possess mountains or valleys, that is, any prominences or cavities.
>
> Thirdly, that it would be absolutely impossible for God to create another world; or even two small bronze spheres at the same time, in the place of these two worlds, as the poles of the parallel axes would coincide because of the lack of an intermediate space.

Nay, even if God could create a world out of these small spheres, closely packed together (disregarding the difficulty of the space that would be left void between them), He would be unable to set them in motion. These are conclusions which Henry More, quite rightly, believed to be indigestible even for a camel's stomach.

Yet Henry More's insistence on the existence of space "outside" the world is, obviously, directed not only against the Aristotelians, but also against the Cartesians to whom he wants to demonstrate the possibility of the limitation of the material world, and at the same time, the mensurability, that is, the existence of dimensions (that now are by no means considered as merely "notional" determinations) in the void space. It seems that More, who in his youth had been such an inspired and enthusiastic adherent of the doctrine of the infinity of the world (and of worlds), became more and more adverse to it, and would have liked to turn back to the "Stoic" conception of a finite world in the midst of an infinite space, or, at least, to join the semi-Cartesians and reject Descartes' infinitization of the material world. He even goes so far as to quote, *with approval,* the Cartesian distinction of the indefiniteness of the world and the infinity of God; interpreting it, of course, as meaning the real *finiteness* of the world opposed to the infinity of space. This, obviously, because he understands now much better than twenty years previously the positive reason of the Cartesian distinction: infinity implies necessity, an infinite *world* would be a necessary one. . . .

But we must not anticipate. Let us turn to another sect of philosophers who are at the same time More's enemies and allies.[21]

> But also those philosophers who did not believe in the creation of matter nevertheless acknowleged [the existence of] Space, such are *Leucippus, Democritus, Demetrius, Metrodorus, Epicurus* and also all the *Stoics.* Some people add Plato to these. As for Aristotle, who defined place (*Locus*) as the nearest surface of the ambient body, he was in this question deserted by a great number of his disciples who rightly observed that in this case he was not in agreement with himself, as indeed he attributed to *place* properties that could not pertain to any thing but to the space occupied by any body; that is, *Equality* and *Immobility.*
>
> It is, moreover, worth while mentioning that those philosophers who made the world finite (such as Plato, Aristotle

and the Stoics) acknowledged *Space* outside the world, or beyond it, whereas those who [believe in] infinite worlds and infinite matter, teach that there is even inside the world an intermixed *vacuum*; such are Democritus and all the Ancients who embraced the atomic philosophy, so that it seems to be entirely confirmed by the voice of nature that there is διατημά τι χωριζοῦ, a certain interval or space really distinct from mundane matter. As for the posteriors, this is sufficiently known. Whereas concerning the *Stoics*, *Plutarch* testifies that they did not admit any void inside the world, but an infinite one outside. And *Plato* says in his *Phaedrus* that above the supreme heaven where he places the purest souls, there is a certain *Supracelestial place* (*locus*), not very different from the abode of the blessed of the Theologians.

As the admission of an infinite space seems thus to be, with very few exceptions, a common opinion of mankind, it may appear unnecessary to insist upon it and to make it an object of discussion and demonstration. More explains therefore that [22]

I should assuredly be ashamed to linger so long upon so easy a question if I were not compelled to do it by the great name of Descartes, who fascinates the less prudent to such an extent that they prefer to rave and rage with Descartes, than to yield to most solid arguments if the *Principles of Philosophy* are opposed to them. Among the most important [tenets] that he himself mentions is that one I have so diligently combatted [elsewhere], namely, that not even by Divine virtue could it happen that there should be in the Universe any interval which, in reality, would not be matter or body. Which opinion I have always considered false; now however I impugn it also as impious. And in order that it should not appear as not completely

> overcome, I shall present and reveal all the subterfuges by which the Cartesians want to elude the strength of my demonstrations, and I shall reply to them.

I must confess that Henry More's answers to the "principal means that the Cartesians used in order to evade the strength of the preceding demonstrations" are sometimes of very dubious value. And that "the refutation of *them all*" is, as often as not, no better than some of his arguments.

Henry More, as we know, was a bad physicist, and he did not always understand the precise meaning of the concepts used by Descartes — for instance, that of the relativity of motion. And yet his criticism is extremely interesting and, in the last analysis, just.[23]

> The first way to escape the strength of our Demonstrations is derived from the Cartesian definition of motion which is as follows: [motion is] *in all cases the translation of a body from the vicinity of those bodies which immediately touch it and are considered as at rest, into the vicinity of others.*[24]

From this definition, objects Henry More, it would follow that a small body firmly wedged somewhere between the axis and the circumference of a large rotating cylinder would be at rest, which is obviously false. Moreover, in this case, this small body, though remaining at rest, would be able to come nearer to, or recede from, another body *P*, placed immobile, outside the rotating cylinder. Which is absurd as "it supposes that there can be an approach of one body to another, quiescent, one without local motion."

Henry More concludes therefore: [25]

. . . that the preceding definition is gratuitously set up by Descartes and, because it is opposed to solid demonstrations, it is manifestly false.

More's error is obvious. It is clear that, *if* we accept the Cartesian conception of the relativity of motion, we no longer have any right to speak of bodies as being absolutely " in motion " or " at rest " but have always to add the point or frame of reference in respect to which the said body is to be considered as being at rest or in motion. And that, accordingly, there is no contradiction in stating that the selfsame body may be at rest in respect to its surroundings and in motion in respect to a body placed farther away, or *vice versa.* And yet Henry More is perfectly right: the extension of the relativity of motion to rotation — at least if we do not want to restrict ourselves to pure kinematics and are dealing with real, physical objects — is illegitimate; moreover, the Cartesian definition, with its more than Aristotelian insistence on the vicinity of the points of reference, is wrong and incompatible with the very principle of relativity. It is, by the way, extremely probable that Descartes thought it out not for purely scientific reasons, but in order to escape the necessity of asserting the motion of the earth and to be able to affirm — with his tongue in his cheek — that the earth was at *rest* in its vortex.

It is nearly the same concerning More's second argument against the Cartesian conception of relativity, or, as More calls it, " reciprocity " of motion. He claims [26]

That the Cartesian definition of motion is rather a description of place; and that if motion were reciprocal, its nature would compel one body to move by two contrary motions and even to move and not to move at the same time.

Thus for instance, let us take three bodies, CD, EF and AB, and let EF move towards H, whilst CD moves

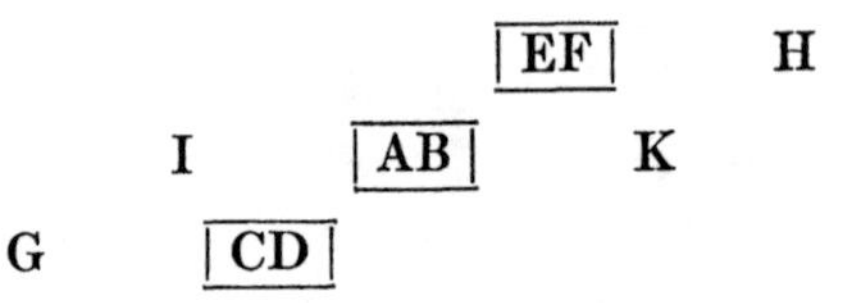

towards G, and AB remains fixed to the earth. Thus it does not move and yet moves at the same time: who can say anything more absurd? And is it not evident [27]

> that the Cartesian definition of motion is repugnant to all the faculties of the soul, the sense, the imagination and the reason.

Henry More, it is clear, cannot transform the concept of motion into that of a pure relation. He feels that when bodies move, even if we consider them as moving in respect to each other, something happens, at least to one of them, that is unilateral and not reciprocal: it *really* moves, that is, changes its place, its internal *locus*. It is in respect to this " place " that motion has to be conceived and not in respect to any other, and therefore [28]

> the supposition of the Cartesians that local motion is relative to the place where the body is not, and not [to the place] where it is, is absurd.

In other terms, relative motion implies absolute motion and can only be understood on the basis of absolute motion and thus of absolute space. Indeed, when a cylindrical body is in circular motion, all its internal points not only change their position in respect to its surrounding surface, or a body placed outside it: they move, that is, pass *through* some extension, describe a trajectory *in* this extension which, therefore, does not move. Bodies do not take their places with them, they go from one place to another. The place of a body, its internal *locus*, is not a part of the body: it is something entirely distinct from it, something that is by no means a mere potentiality of matter: a potentiality cannot be separated from the actual being of a thing, but is an entity, independent of the bodies that are and move in it. And even less is it a mere " phansy," [29] as Dr. Hobbes has tried to assert.

Having thus established, to his own satisfaction, the perfect legitimacy and validity of the concept of space as distinct from matter and refuted their merging together in the Cartesian conception of " extension " Henry More proceeds to the determination of the nature and the ontological status of the corresponding entity.

" Space," or " inner *locus*," is something extended. Now, extension, as the Cartesians are perfectly right in asserting, *cannot be an extension of nothing*: distance between two bodies is something real, or, at the very least, a relation which implies a *fundamentum reale*. The Cartesians, on the other hand, are wrong in believing that void space is nothing. It is something, and even very much so. Once more, it is not a fancy, or a product of imagination, but a perfectly *real* entity. The ancient atomists were right in asserting its reality and calling it an intelligible nature.

The reality of space can be demonstrated also in a somewhat different manner; it is certain [30]

> . . . that a real attribute of any subject can never be found anywhere but where some real subject supports it. But extension is a real attribute of a real subject (namely

> matter), which [attribute] however, is found elsewhere [namely there where no matter is present], and which is independent of our imagination. Indeed we are unable not to conceive that a certain immobile extension pervading everything in infinity has always existed and will exist in all eternity (whether we think about it or do not think about it), and [that it is] nevertheless really distinct from matter.
>
> It is therefore necessary that, because it is a real attribute, some real subject support this extension. This argumentation is so solid that there is none that could be stronger. For if this one fails, we shall not be able to conclude with any certainty the existence in nature of any real subject whatever. Indeed, in this case, it would be possible for real attributes to be present without there being any real subject or substance to support them.

Henry More is perfectly right. On the basis of traditional ontology — and no one in the seventeenth century (except, perhaps, Gassendi, who claims that space and time are neither substances nor attributes but simply space and time) is so bold or so careless as to reject it or to replace it by a new one — his reasoning is utterly unobjectionable. Attributes imply substances. They do not wander alone, free and unattached, in the world. They cannot exist without support, like the grin of the Cheshire cat, for this would mean that they would be attributes *of nothing*. Even those who, like Descartes, modify traditional ontology by asserting that the attributes reveal to us the very nature, or essence, of their substance — Henry More sticks to the old view that they *do not* — maintain the fundamental relationship: no real attribute without real substance. Henry More, therefore, is perfectly right, too, in pointing out that his argumentation is built on exactly the same pattern as the Cartesian and [31]

> . . . that this is the very same means of demonstration as Descartes uses to prove that Space is a substance though it becomes false, in his case, insofar as he concludes that it is a corporeal one.

Moreover, Henry More's conclusion from extension to the underlying and supporting substance is exactly parallel to that of Descartes [31]

> . . . though he [Descartes] aims at another goal than myself. Indeed, from this argument he endeavors to conclude that the Space that is called void is the very same corporeal substance as that called matter. I, on the contrary, since I have so clearly proved that Space or internal place (*locus*) is really distinct from matter, conclude therefrom that it is a certain incorporeal subject or spirit, such as the Pythagoreans once asserted it to be. And so, through that same gate through which the Cartesians want to expel God from the world, I, on the contrary (and I am confident I shall succeed most happily) contend and strive to introduce Him back.

To sum up: Descartes was right in looking for substance to support extension. He was wrong in finding it in matter. The infinite, extended entity that embraces and pervades everything is indeed a substance. But it is not matter. It is Spirit; not *a* spirit, but *the* Spirit, that is, God.

Space, indeed, is not only real, it is something divine. And in order to convince ourselves of its divine character we have only to consider its attributes. Henry More proceeds therefore to the [32]

Enumeration of about twenty titles which the metaphysicians attribute to God and which fit the immobile extended [*entity*] *or internal place* (*locus*).

When we shall have enumerated those names and titles appropriate to it, this infinite, immobile, extended [entity] will appear to be not only something real (as we have just pointed out) but even something Divine (which so certainly is found in nature); this will give us further assurance that it cannot be nothing since that to which so many and such magnificent attributes pertain cannot be nothing. Of this kind are the following, which metaphysicians attribute particularly to the First Being, such as: *One, Simple, Immobile, Eternal, Complete, Independent, Existing in itself, Subsisting by itself, Incorruptible, Necessary, Immense, Uncreated, Uncircumscribed, Incomprehensible, Omnipresent, Incorporeal, All-penetrating, All-embracing, Being by its essence, Actual Being, Pure Act.*

There are not less than twenty titles by which the Divine Numen is wont to be designated, and which perfectly fit this infinite internal place (*locus*) the existence of which in nature we have demonstrated; omitting moreover that the very Divine Numen is called, by the Cabalists, MAKOM, that is, Place (*locus*). Indeed it would be astonishing and a kind of prodigy if the thing about which so much can be said proved to be a mere nothing.

Indeed, it would be extremely astonishing if an entity eternal, uncreated, and existing in and by itself should finally resolve into pure nothing. This impression will only be strengthened by the analysis of the "titles" enumerated by More, who proceeds to examine them one by one: [33]

How this infinite extended [*entity*] *distinct from matter is One, Simple, and Immovable.*

But let us consider the individual titles and note their congruence. This Infinite Extended [entity] distinct from matter is justly called *One*, not only because it is something homogeneous and everywhere similar to itself, but because it is to such an extent one, that it is absolutely impossible that of this one there be many, or that it become many, as it has no physical parts out of which it could be multiplied or in which, truly and physically, it could be divided, or in which it could be condensed. Such indeed is the internal, or, if you prefer, innermost *locus*. From which it follows that it is aptly called *Simple*, since, as I have said, it has no physical parts. As for what pertains to those diversities of which a logical distribution can be made, there is absolutely no thing so simple that they would not be found in it.

But from the Simplicity its Immobility is easily deduced. For no Infinite Extended [entity] which is not co-augmented from parts, or in any way condensed or compressed, can be moved, either part by part, or the whole [of it] at the same time, as it is infinite, nor [can it be] contracted into a lesser space, as it is never condensed, nor can it abandon its place, since this Infinite is the innermost place of all things, inside or outside which there is nothing. And from the very fact that something is conceived as being moved, it is at once understood that it cannot be any part of this Infinite Extended [entity] of which we are speaking. It is therefore necessary that it be immovable. Which attribute of the First Being Aristotle celebrates as the highest.

Absolute space is infinite, immovable, homogeneous, indivisible and unique. These are very important properties which Spinoza and Malebranche discovered almost at the same time as More, and which enabled them to put extension — an intelligible extension, different from that which is given to our imagination and senses — into their respective Gods; properties that Kant — who, however, with Descartes, missed the indivisibility — was to rediscover a hundred years later, and who, accordingly, was unable to connect space with God and had to put it into ourselves.

But we must not wander away from our subject. Let us come back to More, and More's space.[34]

> It is indeed justly called *Eternal*, because we can in no way conceive but that this One, Immovable and Simple [entity] was always, and will be always. But this is not the case for the movable, or for what has physical parts, and is condensed or compressed into parts. Accordingly, Eternity, at least the necessary one, implies also the perfect simplicity of the thing.

We see it at once: space is eternal and therefore uncreated. But the things that are in space by no means participate in these properties. Quite the contrary: they are temporal and mutable and are created by God in the eternal space and at a certain moment of the eternal time.

Space is not only eternal, simple and one. It is also [35]

> . . . *Complete* because it does not coalesce with any other thing in order to form one entity [with it]; otherwise it would move with it at the same time as [that thing], which is not the case of the eternal *locus*.
>
> It is indeed not only Eternal but also *Independent*, not only of our *Imagination*, as we have demonstrated, but of anything whatever, and it is not connected with any other thing or supported by any, but receives and supports all [things] as their site and place.
>
> It must be conceived as *Existing by itself* because it is totally independent of any other. But of the fact that it

> does not depend on anything there is a very manifest sign, namely, that whereas we can conceive all other things as destructible in reality, this Infinite Immovable Extended [entity] cannot be conceived or imagined destructible.

Indeed, we cannot "disimagine" space or think it away. We can imagine, or think of, the disappearance of any object *from* space; we cannot imagine, or think of, the disappearance of space itself. It is the necessary presupposition of our thinking about the existence or non-existence of any thing whatever.[36]

> But that it is *Immense* and *Uncircumscribed* is patent, because wherever we might want to imagine an end to it, we cannot but conceive an ulterior extension which exceeds these ends, and so on *in infinitum.*
>
> Herefrom we perceive that it is incomprehensible. How indeed could a finite mind comprehend that which is not comprehended by any limit?

Henry More could have told us, here too, that he was using, though of course for a different end, the famous arguments by which Descartes endeavoured to prove the indefinity of material extension. Yet he may have felt that not only the goal of the argument, but also its very meaning, opposed it to that of Descartes. Indeed, the *progressus in infinitum* was used by Henry More not for *denying*, but for *asserting* the absolute infinity of the extended substance, which [37]

> . . . is also uncreated, because it is the first of all, for it is by itself (*a se*) and independent of anything else. And *Omnipresent* because it is immense or infinite. But *Incorporeal* because it penetrates matter, though it is a substance, that is, an in-itself subsisting being.

Furthermore it is *All-pervading* because it is a certain immense, incorporeal [entity], and it embraces all the singular [things] in its immensity.

It is even not undeservedly called *Being by essence* in contradistinction to *being by participation*, because *Being by itself* and being *Independent* it does not obtain its essence from any other thing.

Furthermore, it is aptly called *being in act* as it cannot but be conceived as existing outside of its causes.

The list of " attributes " common to God and to space, enumerated by Henry More, is rather impressive; and we cannot but agree that they fit fairly well. After all, this is not surprising: all of them are the formal ontological attributes of the absolute. Yet we have to recognize Henry More's intellectual energy that enabled him not to draw back before the conclusions of his premises; and the courage with which he announced to the world the spatiality of God and the divinity of space.

As for this conclusion, he could not avoid it. Infinity implies necessity. Infinite space is absolute space; even more, it is an Absolute. But there cannot be two (or many) absolute and necessary beings. Thus, as Henry More could not accept the Cartesian solution of the indefiniteness of extension and had to make it infinite, he was *eo ipso* placed before a dilemma: either to make the material world infinite and thus *a se* and *per se*, neither needing, nor even admitting, God's creative action; that is, finally, not needing or even not admitting God's existence at all.

Or he could — and that was exactly what he actually did — separate matter and space, raise the latter to the dignity of an attribute of God, and of an organ in which and through which God creates and maintains His world, a finite world, limited in space as well as in time, as an infinite creature is an utterly contradictory concept. That is something that Henry More acknowledges not to have recognized in his youth when, seized by some poetic furor, he sang in his *Democritus Platonissans* a hymn to the infinity of the worlds.

To prove the limitation in time is not very difficult: it is sufficient, according to More, to consider that nothing can belong to the past if it did not become " past " after having been " present "; and that nothing can ever be " present " if it did not, before that, belong to the future. It follows therefrom that all past events have, at some time, belonged to the future, that is, that there was a time when all of them were not yet " present," not yet existent, a time when everything was still in the future and when nothing was real.

It is much more difficult to prove the limitation of the spatial extension of the (material) world. Most of the arguments alleged in favor of the finiteness are rather weak. Yet it can be demonstrated that the material world must, or at least can, be terminated, and therefore is not really infinite.

> And, in order not to dissimulate anything, this seems to be the best argument for demonstrating that the Matter of the World cannot be absolutely infinite but only indefinite, as Descartes has said somewhere, and to reserve the name of infinite for God alone. Which must be asserted as well of the *Duration* as of the *Amplitude of God*. Both are indeed absolutely infinite; those of the World, however, only indefinite . . . that is, in truth, finite. In this way God is duly, that is, infinitely, elevated above the Universe, and

> is understood to be not only by an infinite eternity older than the World, but also by immense spaces larger and more ample than it.

The circle is closed. The conception that Henry More ascribed to Descartes — though falsely — and so bitterly criticized in his youth, has demonstrated its good points. An indeterminately vast but finite world merged in an infinite space is the only conception, Henry More sees it now, that enables us to maintain the distinction between the contingent created world and the eternal and *a se* and *per se* existing God.

By a strange irony of history, the *κενόν* of the godless atomists became for Henry More God's own extension, the very condition of His action in the world.

VII. Absolute Space, Absolute Time and Their Relations to God

Malebranche
Newton
& Bentley

Henry More's conception of space, which makes it an attribute of God, is by no means — I have said it already, but I should like to insist upon it — an aberrant, odd and curious invention, a "fancy," of a Neoplatonic mystic lost in the world of the new science. Quite the contrary. It is, in its fundamental features, shared by a number of the great thinkers of his time, precisely those who identified themselves with the new scientific world-view.

I need not insist on Spinoza who, though he denied the existence of void space and maintained the Cartesian identification of extension and matter, carefully distinguishes between extension, as given to the senses and represented by the imagination, and extension as perceived by the understanding — the former, being divisible and movable (and corresponding to the Cartesian indefinitely extended world), constituting the sempiternal many-fold of ever-changing and finite *modi*, the latter, truly and fully infinite and therefore indivisible, constituting the eternal and essential attribute of the *a se* and *per se* existing Being, that is, of God.

Infinity belongs unavoidably to God, not only to the very dubious God of Spinoza, but also to the God of the Christian religion. Thus, not only Spinoza, the by no means pious Dutch philosopher, but also the very pious Father Malebranche, having grasped the essential infinity of geometrical space, is obliged to connect it with God. The space of geometers or, as Malebranche calls it, the "intelligible extension," is, according to Christ Himself, who appears as one of the interlocutors of the *Christian Meditations* of Malebranche,[1]

> . . . eternal, immense, necessary. It is the immensity of the Divine Being, as infinitely participable by the corporeal creature, as representative of an immense matter; it is, in a word, the intelligible idea of possible worlds. It is what your mind contemplates when you think about the infinite. It is by means of this intelligible extension that you know the visible world.

Malebranche, of course, does not want to put matter into God and to spatialize God in the manner in which Henry More or Spinoza did it. He distinguishes therefore the *idea* of space, or "intelligible extension," which he places in God, from the gross material extension of the world created by God.[2]

> But you have to distinguish two kinds of extension, the one intelligible, and the other material.

The intelligible extension is "eternal, necessary, infinite," whereas the [3]

> . . . other kind of extension is that which is created; it is

> the matter out of which the world is built. . . . This world began and can cease to be. It has certain limits that it cannot lack. . . . Intelligible extension appears to you eternal, necessary, infinite; believe what you see; but do not believe that the world is eternal, or that the matter that composes it is immense, necessary, eternal. Do not attribute to the creature what pertains only to the Creator, and do not confuse My [Christ's] substance which God engenders by the necessity of His Being with My work which I produce with the Father and the Holy Spirit by an entirely free operation.

It is just the confusion between the intelligible extension and the created one that induces some people to assert the eternity of the world and to deny its creation by God. For,[4]

> there is another reason which leads men to believe that matter is uncreated; indeed, when they think about extension they cannot prevent themselves from regarding it as a necessary being. For they conceive that the world has been created in immense spaces, that these spaces never had a beginning, and that God Himself cannot destroy them. Thus, confusing matter with these spaces, as matter effectively is nothing else but space or extension, they regard matter as an Eternal being.

This is, as a matter of fact, a rather natural error as Malebranche himself does not fail to point out to his Divine Master; he recognizes, of course, that his doubts are removed, and that he now sees the distinction that formerly escaped him. Still [5]

> I beg you, had I not some reason to believe that extension is eternal? Must one not judge things according to one's ideas, and is it even possible to judge otherwise? And,

as I cannot prevent myself from regarding intelligible extension as immense, eternal, necessary, had I not grounds for thinking that material extension has the same attributes?

By no means. In spite of the Cartesian axiom hinted at by Malebranche (in the role of the *discipulus* of the dialogue), according to which we are entitled to assert of the thing what we clearly perceive to belong to its idea, the reasoning attributing infinity and eternity to material extension was illegitimate; thus the Divine Master replies: [6]

> We must, my dear Disciple, judge things according to their ideas; it is only thus that we have to judge them. But that concerns their essential attributes, and not the circumstances of their existence. The idea you have of extension represents it to you as divisible, mobile, impenetrable: judge without fear that it has essentially these properties. But do not judge that it is immense, or that it is eternal. It may not exist at all, or possess very narrow limits. [The contemplation of the idea of extension] gives you no reason to believe that there is [in existence] even one foot of material extension, though you have present in your mind an infinite immensity of intelligible extension; and much less are you entitled to judge that the world is infinite as some philosophers assert. Do not judge either that the world is eternal because you regard intelligible extension as a necessary being of which the duration has no beginning and cannot have an end. For, though you must judge the essence of things according to the ideas which represent them, you must never judge by them of their existence.

The Disciple of Malebranche's dialogue is fully convinced — who, indeed, would not be by such a Master? Nobody else, alas, shared his conviction.

Antoine Arnauld considered the Malebranchian distinction between " intelligible " and " created " extension as perfectly spurious and corresponding only and solely to the Cartesian distinction between (real) extension given to the senses and the same real extension as object of pure understanding. According to him Malebranche's " intelligible extension " was simply the infinite extension of the material universe. Thirty years later, Dortous de Mairan made fundamentally the same reproach, though he formulated it in a somewhat different and much nastier manner: according to him Malebranche's " intelligible extension " was indistinguishable from that of Spinoza. . . .[7]

But not only philosophers shared, more or less, Henry More's conception of space: it was shared by Newton, and this, because of the unrivaled influence of Newton on the whole subsequent development, is, indeed, of overwhelming importance.

It may seem strange, at first glance, to link together Henry More and Isaac Newton. . . . And yet, this link is perfectly established.[8] Moreover, as we shall see, More's explicit teaching will throw some light on the implicit premises of Newtonian thinking, a light all the more necessary as Isaac Newton, in contradistinction not only to Henry More but also to René Descartes, is neither a professional metaphysician like the former, nor, like the latter, at once a great philosopher and a great scientist: he is a professional scientist, and though science, at that time, had not yet accomplished its disastrous divorce from philosophy, and though physics was still not only designated, but also thought of, as " natural philosophy," it is nevertheless true that his primary interests are in the field of " science," and not of " philosophy." He deals, therefore, with metaphysics not *ex professo*, but only insofar as he needs it to establish the foundations of his intentionally empirical and allegedly positivistic mathematical investigation of nature. Thus the metaphysical pronouncements of Newton are not very numerous and, Newton being a very cautious and secretive person as well as a very careful writer, they are rather reticent and reserved. And yet they are sufficiently clear so as not to be misunderstood by his contemporaries.

Newton's physics, or, it would be better to say, Newton's natural philosophy, stands or falls with the concepts of absolute time and absolute space, the selfsame concepts for which Henry More fought his long-drawn-out and relentless battle against Descartes. Curiously enough, the Cartesian conception of the only relative, or relational, character of these and connected notions is branded by Newton as being " vulgar " and as based upon " prejudices."

Thus in the famous *scholium* which follows the *Definitions* that are placed at the very beginning of the *Principia*, Newton writes: [9]

> Hitherto I have laid down the definitions of such words as are less known, and explain the sense in which I would have them to be understood in the following discourse. I do not define time, space, place, and motion as being well known to all. Only I must observe that the vulgar conceive those quantities under no other notions but from the relations they bear to sensible objects. And thence arise certain prejudices, for the removing of which, it will be convenient to distinguish them into absolute and relative, true and apparent, mathematical and common.

Absolute, true and mathematical time and space — for Newton these qualifications are equivalent and determine the nature both of the concepts in question and of the entities corresponding to them — are thus, in a manner of which we have already seen some examples, *opposed* to the merely common-sense time and space. As a matter of fact, they could just as well be called " intelligible " time and space in contradistinction to " sensible." Indeed, according to the " empiricist " Newton,[10] " in philosophical disquisitions we ought to abstract from our senses and consider things themselves, distinct from what are only sensible measures of them." Thus: [11]

> It may be that there is no such thing as an equable motion whereby time may be accurately measured. All motion may be accelerated and retarded, but the flowing of absolute time is liable to no change. The duration or perseverance of the existence of things remains the same; whether the motions are swift or slow, or none at all: and therefore it ought to be distinguished from what are only sensible measures thereof.

Time is not only not linked with motion — like Henry More before him, Newton takes up against Aristotle the Neoplatonic position — it is a reality in its own right: [12]

> Absolute, true and mathematical time, of itself and from its own nature, flows equably without regard to anything external,

that is, it is *not*, as Descartes wants us to believe, something which pertains only to the external, material world and which would not exist if there were no such world, but something which has its *own nature* (a rather equivocal and dangerous assertion which Newton later had to correct by relating time, as well as space, to God), " and by another name is called duration "; that is, once more, time is *not*, as Descartes wants us to believe, something subjective and distinct from the duration which he, Descartes, identifies with the amount of reality of the created being. Time and duration are only two names for the same objective and absolute entity.

But, of course,[13]

> . . . relative, apparent and common time, is some sensible and external (whether accurate or unequable) measure of duration by the means of motion, which is commonly used instead of true time: such as an hour, a day, a month, a year.

It is just the same concerning space: [14]

> Absolute space, in its own nature, without regard to anything external, remains always similar and immovable,

that is, space is *not* Cartesian extension which moves around, and which by Descartes is identified with, bodies. This is, at most, *relative* space, which is mistaken for the absolute space that subtends it by both Cartesians and Aristotelians.[15]

> Relative space is some movable dimension or measure of the absolute spaces; which our senses determine by its position to bodies, and which is vulgarly taken for immovable space; such is the dimension of a subterraneous, an aereal, or celestial space, determined by its position in respect of the earth. Absolute and relative space are the same in figure and magnitude; but they do not remain always numerically the same,

because relative space, which is, so to speak, attached to the body, moves with that body through absolute space.[16]

> For if the earth, for instance, moves, a space of our air, which relatively and in respect of the earth always remains

the same, will at one time be one part of the absolute space into which the air passes; and another time will be another part of the same and so, absolutely understood, it will be perpetually mutable.

Just as we have distinguished absolute, immovable space from the relative spaces that are and move in it, so we have to make a distinction between absolute and relative *places* which bodies occupy in space. Thus, elaborating More's analysis of this concept and his criticism of the traditional as well as the Cartesian conceptions, Newton states: [17]

> Place is a part of space which a body takes up and is, according to the space, either absolute or relative. I say, a part of space; not the situation nor the external surface of the body. For the places of equal solids are always equal; but their surfaces, by reason of their dissimilar figures, are often unequal. Positions properly have no quantity; nor are they so much the places themselves as the properties of places. The motion of the whole is the same with the sum of the motions of the parts; that is, the translation of the whole, out of its place, is the same thing with the sum of the translations of the parts out of their places; and therefore the place of the whole is the same as the sum of the places of the parts, and for that reason it is internal and in the whole body.

Place — *locus* — is thus something which is *in* the bodies, and *in which* bodies are in their turn. And as motion is a process in which bodies change their places, not taking them along with them but relinquishing them for others, the distinction between absolute and relative spaces implies necessarily that of absolute and relative motion, and *vice versa*, is implied by it: [18]

Absolute motion is the translation of a body from one absolute place into another, and relative motion the translation from one relative place into another. Thus in a ship under sail the relative place of a body is that part of the ship which the body possesses, or that part of the cavity which the body fills and which therefore moves together with the ship, and relative rest is the continuance of the body in the same part of the ship or of its cavity. But real, absolute rest is the continuance of the body in the same part of that immovable space in which the ship itself, its cavity, and all that it contains is moved. Wherefore, if the ship is really at rest, the body, which relatively rests in the ship, will really and absolutely move with the same velocity which the ship has on the earth. But if the earth also moves, the true and absolute motion of the body will arise, partly from the true motion of the earth in immovable space, partly from the relative motion of the ship on the earth; and if the body moves also relatively in the ship, its true motion will arise, partly from the true motion of the earth in immovable space and partly from the relative motions as well of the ship on the earth as of the body in the ship; and from these relative motions will arise the relative motion of the body on the earth. As if that part of the earth where the ship is was truly moved toward the east with a velocity of 10,000 parts, while the ship itself, with a fresh gale and full sails, is carried toward the west with a velocity expressed by 10 of those parts, but a sailor walks in the ship toward the east with 1 part of the said velocity; then the sailor will be moved truly in immovable space toward the east, with a velocity of 10,001 parts, and relatively on the earth toward the west, with a velocity of 9 of those parts.

As for the inner structure of space, it is characterized by Newton in terms that strongly remind us of the analysis made by Henry More: [19]

As the order of the parts of time is immutable, so also is the order of the parts of space. Suppose those parts to be moved out of their places, and they will be moved (if the expression may be allowed) out of themselves. For times and spaces are, as it were, the places as well of themselves as of all other things. All things are placed in time as to order of succession and in space as to order of situation. It is from their essence or nature that they are places, and that the primary places of things should be movable is absurd. These are therefore the absolute places, and translations out of those places are the only absolute motions.

Newton, it is true, does not tell us that space is " indivisible " or " indiscerpible "; [20] yet it is obvious that to " divide " Newton's space, that is, actually and really to separate its " parts," is just as impossible as it is impossible to do so with More's, an impossibility that does not preclude the making of " abstract " or " logical " distinctions and divisions, or prevent us from distinguishing inseparable " parts " in absolute space and from asserting its indefinite, or even infinite " divisibility." Indeed, for Henry More, as well as for Newton, the infinity and the continuity of absolute space imply the one as well as the other.

Absolute motion is motion in respect to absolute space, and all relative motions imply absolute ones: [21]

. . . all motions, from places in motion, are no other than parts of entire and absolute motions; and every entire motion is composed of the motion of the body out of its first place and the motion of this place out of its place; and so on, until we come to some immovable place, as in the before-mentioned example of the sailor. Wherefore entire and absolute motions cannot be otherwise determined than

> by immovable places; and for that reason I did before refer those absolute motions to immovable places, but relative ones to movable places. Now no other places are immovable but those that, from infinity to infinity, do all retain the same given position one to another, and upon this account must ever remain unmoved and do thereby constitute immovable space.

"*From infinity to infinity* retain the same position. . . ." What does *infinity* mean in this place? Obviously not only the spatial, but also the temporal: absolute places retain from *eternity to eternity* their positions in the absolute, that is, *infinite* and *eternal* space, and it is in respect to this space that the motion of a body is defined as being absolute.

Alas, absolute motion is very difficult, or even impossible, to determine. We do not perceive space — it is, as we know, inaccessible to our senses. We perceive things in space, their motions in respect to other things, that is, their relative motions, not their absolute motions in respect to space itself. Moreover, motion itself, or in itself, the *status* of motion, though utterly opposed to the *status* of rest, is nevertheless (as we see it clearly in the fundamental case of uniform, rectilinear, *inertial* motion) absolutely indistinguishable from the latter.

It is only by their causes and effects that absolute and relative motions can be distinguished and determined: [22]

> The causes by which true and relative motions are distinguished, one from the other, are the forces impressed upon bodies to generate motion. True motion is neither generated nor altered but by some force impressed upon the body moved, but relative motion may be generated or altered without any force impressed upon the body. For it

> is sufficient only to impress some force on other bodies with which the former is compared that, by their giving way, that relation may be changed in which the relative rest or motion of this other body did consist. Again, true motion suffers always some change from any force impressed upon the moving body, but relative motion does not necessarily undergo any change by such forces. For if the same forces are likewise impressed on those other bodies with which the comparison is made, that the relative position may be preserved, then that condition will be preserved in which the relative motion consists. And therefore any relative motion may be changed when the true motion remains unaltered, and the relative may be preserved when the true suffers some change. Thus, true motion by no means consists in such relations.

Thus it is only in the cases where our determination of the forces acting upon the bodies is not based upon the perception of the change of the mutual relations of the bodies in question that we are actually able to distinguish absolute motions from relative ones, or even from rest. Rectilinear motion, as we know, does not offer us this possibility. But circular or rotational motion does.[23]

> The effects which distinguish absolute from relative motion are the forces of receding from the axis of circular motion. For there are no such forces in a circular motion purely relative, but in a true and absolute circular motion they are greater or less, according to the quantity of the motion.

Rotational or circular motion, everywhere on the earth as in the skies, gives birth to centrifugal forces, the determination of which enables us to recognize its existence in a given body, and even to measure its speed, without taking into account the positions or behavior of any other body outside the gyrating one. The purely relative conception finds its limit — and its refutation — in the case of circular motion and, at the same time, the Cartesian endeavor to extend this conception to celestial motions appears as it really is: a clumsy attempt to disregard the facts, a gross misinterpretation or misrepresentation of the structure of the universe.[24]

> There is only one real circular motion of any one revolving body, corresponding to only one power of endeavoring to recede from its axis of motion, as its proper and adequate effect; but relative motions, in one and the same body, are innumerable, according to the various relations it bears to external bodies, and, like other relations, are altogether destitute of any real effect, any otherwise than they may perhaps partake of that one only true motion. And therefore in their system who suppose that our heavens, revolving below the sphere of the fixed stars, carry the planets along with them, the several parts of those heavens and the planets, which are indeed relatively at rest in their heavens, do yet really move. For they change their position one to another (which never happens to bodies truly at rest) and, being carried together with their heavens, partake of their motions and, as parts of revolving wholes, endeavour to recede from the axis of their motions.

The Newtonian discovery of the absolute character of rotation — in contradistinction to rectilinear translation — constitutes a decisive confirmation of his conception of space; it makes it accessible to our empirical knowledge and, without depriving it of its metaphysical function and *status*, it ensures its role and its place as a fundamental concept of science.

The Newtonian interpretation of circular motion as

motion " relative " to absolute space, and, of course, the very idea of absolute space with its physico-metaphysical implications, met, as we know, with rather strong opposition. For two hundred years, from the times of Huygens and Leibniz to those of Mach and Duhem, it was subjected to searching and vigorous criticism.[24a] It has, in my opinion, withstood victoriously all the assaults, which is, by the way, not so very surprising: it is indeed the necessary and inevitable consequence of the " bursting of the sphere," the " breaking of the circle," the geometrization of space, of the discovery or assertion of the law of inertia as the first and foremost law or axiom of motion. Indeed, if it is the inertial, that is, the rectilinear uniform motion that becomes — just like rest — the " natural " *status* of a body, then the circular one, which at any point of its trajectory *changes* its direction though maintaining constant its angular velocity, appears, from the point of view of the law of inertia, not as a *uniform*, but as a *constantly accelerated* motion. But acceleration, in contradistinction to mere translation, has always been something absolute, and it remained so until 1915, when, for the first time in the history of physics, the general relativity theory of Einstein deprived it of its absoluteness. Yet as, in so doing, it reclosed the universe and denied the Euclidean structure of space, it has, by this very fact, confirmed the correctness of the Newtonian conception.

Newton thus was perfectly right in stating that we are able to determine the absolute rotational or circular motion of bodies without needing, for that purpose, a term of reference represented by a body at absolute rest; though he was wrong, of course, in his pious hope of being able, finally, to achieve the determination of all "true" motions. The difficulties that stood in his path were not merely — as he believed them to be — very great. They were insurmountable.[25]

> It is indeed a matter of great difficulty to discover and effectively to distinguish the true motions of particular bodies from the apparent, because the parts of that immovable space in which those motions are performed do by no means come under the observation of our senses. Yet the thing is not altogether desperate; for we have some arguments to guide us, partly from the apparent motions, which are the differences of the true motions; partly from the forces, which are the causes and effects of the true motions. For instance, if two globes, kept at a given distance one from the other by means of a cord that connects them, were revolved about their common center of gravity, we might, from the tension of the cord, discover the endeavor of the globes to recede from the axis of their motion, and from thence we might compute the quantity of their circular motions. And then if any equal forces should be impressed at once on the alternate faces of the globe to augment or diminish their circular motions, from the increase or decrease of the tension of the cord we might infer the increment or decrement of their motions, and thence would be found on what faces those forces ought to be impressed that the motions of the globes might be most augmented; that is, we might discover their hindmost faces, or those which, in the circular motion, do follow. But the faces which follow being known, and consequently the opposite ones that precede, we should likewise know the determination of their motions. And thus we might find the quantity and the determination of this circular motion, even in an immense vacuum, where there was nothing external or

sensible with which the globes could be compared. But now, if in that space some remote bodies were placed that kept always a given position one to another, as the fixed stars do in our regions, we could not indeed determine from the relative translation of the globes among those bodies whether the motion did belong to the globes or to the bodies. But if we observed the cord and found that its tension was that very tension which the motion of the globes required, we might conclude the motion to be in the globes and the bodies to be at rest; and then, lastly, from the translation of the globes among the bodies, we should find the determination of their motions. But how we are to obtain the true motions from their causes, effects, and apparent differences, and the converse, shall be explained at large in the following treatise. For to this end it was that I composed it.

The real distinction between space and matter, though it involves the rejection of the Cartesian identification of the essence of matter with extension, does not, as we know, necessarily imply the acceptance of the existence of an actual vacuum: we have seen Bruno, and Kepler too, assert that space is everywhere full of "ether." As for Newton, though he, too, believes in an ether that fills at least the space of our "world" (solar system), his ether is only a very thin and very elastic substance, a kind of exceedingly rare gas, and it does not completely fill the world space. It does not extend itself to infinity as is sufficiently clear from the motion of comets:[26]

. . . for though they are carried in oblique paths and sometimes contrary to the course of the planets, yet they move every way with the greatest freedom, and preserve their motion for an exceeding long time, even when contrary

> to the course of the planets. Hence also it is evident that the celestial spaces are void of resistance,

and as unresisting matter, that is, matter deprived of the *vis inertiae,* is unthinkable, it is obvious that the celestial spaces are void also of matter. Moreover, even where it is present, Newtonian ether does not possess a continuous structure. It is composed of exceedingly small particles between which, of course, there is vacuum. Elasticity, indeed, implies vacuum. In a Cartesian world, that is, in a world constituted by a continuously-spread uniform matter, elasticity would be impossible. Nay, if all spaces were equally full (as they must be according to Descartes) even motion would not be possible.[27]

> All spaces are not equally full; for if all spaces were equally full, then the specific gravity of the fluid which fills the region of the air, on account of the extreme density of the matter, would fall nothing short of the specific gravity of quicksilver, or gold, or any other the most dense body; and, therefore, neither gold nor any other body could descend in air; for bodies do not descend in fluids, unless they are specifically heavier than the fluids. And if the quantity of matter in a given space can, by any rarefaction, be diminished, what should hinder a diminution to infinity?

Matter, according to Newton, who shares the atomic conceptions of his contemporaries (and even improves upon them in a very interesting manner), has an essentially granular structure. It is composed of small, solid, particles and therefore [28]

> if all the solid particles of all bodies are of the same density and cannot be rarefied without pores, then a void space, or vacuum, must be granted.

As for matter itself, the essential properties that Newton ascribes to it are nearly the same as those that have been listed by Henry More,[29] by the old atomists and the modern partisans of corpuscular philosophy: extension, hardness, impenetrability, mobility, to which is added — a most important addition — inertia, in the precise, new meaning of this word. In a curious combination of anti-Cartesian empiricism and ontological rationalism, Newton wants to admit as *essential* properties of matter only those that are (a) empirically given to us, and (b) can be neither increased nor diminished. Thus he writes in the third of his *Rules of Reasoning in Philosophy*, by which he replaced the third fundamental *Hypothesis* of the first edition of the *Principia:* [30]

> *The qualities of bodies, which admit neither intensification nor remission of degrees, and which are found to belong to all bodies within the reach of our experiments, are to be esteemed the universal qualities of all bodies whatsoever.*
>
> For since the qualities of bodies are only known to us by experiments, we are to hold for universal all such as universally agree with experiments, and such as are not liable to diminution can never be quite taken away. We are certainly not to relinquish the evidence of experiments for the sake of dreams and vain fictions of our own devising; nor are we to recede from the analogy of Nature, which is wont to be simple and always consonant to itself. We in no other way know the extension of bodies than by our senses, nor do these reach it in all bodies; but because we perceive extension in all that are sensible, therefore we ascribe it universally to all others also. That abundance of bodies are hard we learn by experience; and because the hardness of the whole arises from the hardness of the parts, we therefore justly infer the hardness of the undivided

particles, not only of the bodies we feel, but of all others. That all bodies are impenetrable, we gather not from reason, but from sensation. The bodies which we handle we find impenetrable, and hence conclude impenetrability to be a universal property of all bodies whatsoever. That all bodies are movable and endowed with certain powers (which we call the inertia) of persevering in their motion, or in their rest, we only infer from the like properties observed in the bodies which we have seen. The extension, hardness, impenetrability, mobility, and inertia of the whole result from the extension, hardness, impenetrability, mobility, and inertia of the parts and hence we conclude the least particles of all bodies to be also all extended, and hard and impenetrable, and movable, and endowed with their proper inertia. And this is the foundation of all philosophy. Moreover, that the divided but contiguous particles of bodies may be separated from one another is a matter of observation; and, in the particles that remain undivided, our minds are able to distinguish yet lesser parts, as is mathematically demonstrated. But whether the parts so distinguished and not yet divided may, by the powers of Nature, be actually divided and separated from one another we cannot certainly determine. Yet had we the proof of but one experiment that any undivided particle, in breaking a hard and solid body, suffered a division, we might by virtue of this rule conclude that the undivided as well as the divided particles may be divided and actually separated to infinity.

Lastly, if it universally appears, by experiments and astronomical observations, that all bodies about the earth gravitate toward the earth, and that in proportion to the quantity of matter which they severally contain; that the moon likewise, according to the quantity of its matter, gravitates toward the earth; that, on the other hand, our sea gravitates toward the moon; and all the planets one toward another; and the comets in like manner toward the sun: we must, in consequence of this rule, universally allow that all bodies whatsoever are endowed with a principle of mutual gravitation. For the argument from the appearances concludes with more force for the universal gravitation of all bodies than for their impenetrability, of which, among those in the celestial regions, we have no experiments nor any manner of observation. Not that I affirm gravity to be essential to bodies; by their *vis insita* I mean nothing but their inertia. This is immutable. Their gravity is diminished as they recede from the earth.

We see, therefore, that Newton, no more than Galileo or even Descartes, includes gravity, or mutual attraction, in the essential properties of bodies in spite of the fact that its empirical foundations are much stronger than those of such a fundamental property as impenetrability. Newton seems to suggest that the reason for this exclusion consists in the variability of gravitation as opposed to the immutability of the inertia. But this is by no means the case. The *weight* of a body "gravitating" toward the earth is indeed diminished as it recedes from it: but the attractive force of the earth — or any other body — is constant, and, just as in the case of inertia, proportional to its mass, and it is as such that it appears in the famous inverse square formula of universal gravitation. This is so because [31]

. . . it is reasonable to suppose that forces which are directed to bodies should depend upon the nature and quantity of those bodies, as we see they do in magnetical experiments. And when such cases occur, we are to compute the attractions of the bodies by assigning to each of their particles its proper force, and then finding the sum of them all.

Thus the attraction of a body is a function, or sum, of the attractions of its (atomic) particles, just as its mass is the sum of the masses of the selfsame particles. And yet it is not an " essential property " of the body, or of its particles. As a matter of fact it is not even an adventitious property of them; it is in no sense their property. It is an effect of some extraneous force acting upon them according to a fixed rule.

It is — or should be — a well-known fact that Newton did not believe in attraction as a real, physical force. No more than Descartes, Huygens or Henry More could he admit that matter is able to act at a distance, or be animated by a spontaneous tendency. The empirical corroboration of the fact could not prevail against the rational impossibility of the process. Thus, just like Descartes or Huygens, he tried at first to explain attraction — or to explain it away — by reducing it to some kind of effect of purely mechanical occurrences and forces. But in contradistinction to the former, who believed that they were able to devise a mechanical theory of gravity, Newton seems to have become convinced of the utter futility of such an attempt. He discovered, for example, that he could indeed explain attraction, but that in order to do so he had to postulate repulsion, which, perhaps, was somewhat better, but not very much so.

Fortunately, as Newton knew full well, we need not have a clear conception of the way in which certain effects are produced in order to be able to study the phenomena and to treat them mathematically. Galileo was not obliged to develop a theory of gravity — he even claimed his right to ignore completely its nature — in order to establish a mathematical dynamics and to determine the laws of fall.[32]

Thus nothing prevented Newton from studying the *laws* of " attraction " or " gravitation " without being obliged to give an account of the real forces that produced the centripetal motion of the bodies. It was perfectly sufficient to assume only that these forces — whether physical or metaphysical — were acting according to strict mathematical laws (an assumption fully confirmed by the observation of astronomical phenomena and also by well-interpreted experiments) and to treat these " forces " as *mathematical* forces, and not as real ones. Although only part of the task, it is a very necessary part; only when this preliminary stage is accomplished can we proceed to the investigation of the real causes of the phenomena.

This is precisely what Newton does in the book so significantly called not *Principia Philosophiae*, that is, *Principles of Philosophy* (like Descartes'), but *Philosophiae naturalis principia mathematica*, that is, MATHEMATICAL *Principles of* NATURAL *Philosophy*. He warns us that: [33]

> I here use the word " attraction " in general for any endeavor whatever made by bodies to approach each other, whether that endeavor arise from the action of the bodies themselves, as tending to each other or agitating each other by spirits emitted; or whether it arises from the action of the ether or of the air, or of any medium whatever, whether corporeal or incorporeal, in any manner impelling bodies placed therein toward each other. In the same general sense I use the word *impulse*, not defining in this treatise the species or physical qualities of forces, but investigating the quantities and mathematical proportions of them, as I observed before in the definitions. In mathematics we are to investigate the quantities of forces with their proportions consequent upon any conditions supposed; then, when we

enter upon physics, we compare these proportions with the phenomena of Nature, that we may know what conditions of these forces answer to the several kinds of attractive bodies. And this preparation being made, we argue more safely concerning the physical species, causes, and proportion of the forces.

In his *Letters* (written five years after the publication of the *Principia*) to Richard Bentley who, like nearly everybody else, missed the warning just quoted and interpreted Newton in the way that became common in the eighteenth century, namely as asserting the *physical* reality of attraction and of attractive force as inherent to matter, Newton is somewhat less reserved. He first tells Bentley (in his second letter): [34]

You sometimes speak of gravity as essential and inherent to matter. Pray do not ascribe that notion to me, for the cause of gravity is what I do not pretend to know and therefore would take more time to consider of it.

In the third one, he practically comes into the open. Though he does not tell Bentley what he, Newton, believes the force of attraction to be *in rerum*, he tells him that: [35]

It is inconceivable that inanimate brute matter should, without mediation of something else which is not material, operate upon and affect other matter without mutual contact, as it must be if gravitation, in the sense of Epicurus, be essential and inherent in it. And this is one reason why I desired you would not ascribe innate gravity to me. That gravity should be innate, inherent, and essential to matter, so that one body may act upon another at a distance through a *vacuum*, without the mediation of anything else, by and through which their action and force may be conveyed from one to another, is to me so great an absurdity that I believe no man who has in philosophical matters a competent faculty of thinking can ever fall into it. Gravity must be caused by an agent acting constantly according to certain laws, but whether this agent be material or immaterial I have left to the consideration of my readers.

As we see, Newton does *not* pretend any longer not to *know* the cause of gravity; he only informs us that he left this question unanswered, leaving it to his readers to find out themselves the solution, namely that the "agent" which "causes" gravity cannot be material, but must be a spirit, that is, either the spirit of nature of his colleague Henry More, or, more simply, God — a solution that, rightly or wrongly, Newton was too cautious to announce himself. But that Dr. Bentley could not — and did not — fail to understand.

As for Dr. Bentley (or more exactly Mr. Richard Bentley, M.A. — he became DD. in 1696), who did not know much physics — he was by training a classicist — and obviously did not grasp the ultimate implications of Newton's natural philosophy, he espouses it wholeheartedly, as far, at least, as he understands it, and turns it into a weapon for the *Confutation of Atheism* in the Boyle Lectures which he gave in 1692.

Richard Bentley follows so closely, and even so servilely, Newton's teaching, or lessons — he copied nearly *verbatim* the letters he received from him, adding, of course, some references to the Scriptures and a good deal of rhetoric — that the views he expresses can be considered as representing, in a large measure, those of Newton himself.

The atheists Mr. Bentley deals with are essentially the

materialists, more precisely those of the Epicurean brand, and it is rather amusing to see that Bentley accepts the fundamentals of their conception, that is, the corpuscular theory of matter, the reduction of material being to atoms and void, not only without the apparent hesitations and cautious reserve of Newton, but even as something that goes without saying and without discussion. He only objects, as it has always been done, that it is not enough, and that they cannot explain the orderly structure of our universe without superadding to matter and motion some purposeful action of a non-material cause: the fortuitous and disorderly motion of atoms cannot transform chaos into a cosmos.

Yet, if the pattern of his reasoning is quite traditional — but we must not blame Mr. Bentley for that: it is also the Newtonian pattern and, moreover, did not Kant himself tell us a century later that the physico-teleological proof of the existence of God is the only one that has any value? — the contents of the demonstration are adapted to the present-day (Bentley's present day) level of scientific philosophy.

Thus, for instance, he accepts without the slightest criticism the contemporary version of Giordano Bruno's conception of the universe: an infinite space with an immense number of star-suns. Bentley maintains, of course, that the stars are finite in number — he thinks he can prove it — and would even like them to be arranged in space so as to build a "firmament." But if this cannot be done, he will accept their dispersion in the boundless void. Bentley, indeed, insists upon the void. He needs it, of course, as we shall see in a moment, in order to be able to demonstrate the existence and the action, in the world, of non-material, non-mechanical forces — first and foremost of the Newtonian universal attraction — but he is also somehow elated and ravished by the idea that this our world is chiefly composed of void spaces, and he indulges in calculations that show that the amount of matter in the universe is so small as practically not to be worth speaking of: [36]

> Let us allow, then, that all the matter of the system of our sun may be 50,000 times as much as the whole mass of the earth; and we appeal to astronomy, if we are not liberal enough and even prodigal in this concession. And let us suppose further, that the whole globe of the earth is entirely solid and compact, without any void interstices; notwithstanding what hath been shewed before, as to the texture of gold itself. Now, though we have made such ample allowances, we shall find, notwithstanding, that the void space of our system is immensely bigger than all its corporeal mass. For, to proceed upon supposition, that all the matter within the firmament is 50,000 times bigger than the solid globe of the earth; if we assume the diameter of the *orbis magnus* (wherein the earth moves about the sun) to be only 7,000 times as big as the diameter of the earth, (though the latest and most accurate observations make it thrice 7,000), and the diameter of the firmament to be only 100,000 times as long as the diameter of the *orbis magnus* (though it cannot possibly be less than that, but may be vastly and unspeakably bigger), we must pronounce, after such large concessions on that side, and such great abatements on ours, that the sum of empty spaces within the concave of the firmament is 6,860 million million million times bigger than all the matter contained in it.
>
> . . . .
>
> And first, because every fixed star is supposed by astron-

omers to be of the same nature with our sun, and each may very possibly have planets about them, though, by reason of their vast distance, they may be invisible to us; we will assume this reasonable supposition, that the same proportion of void space to matter, which is found in our sun's region within the sphere of the fixed stars, may competently well hold in the whole mundane space. I am aware that in this computation we must not assign the whole capacity of that sphere for the region of our sun, but allow half of its diameter for the *radii* of the several regions of the next fixed stars; so that, diminishing our former number, as this last consideration requires, we may safely affirm, from certain and demonstrated principles, that the empty space of our solar region (comprehending half of the diameter of the firmament) is 8,575 hundred thousand million million times more ample than all the corporeal substance in it. And we may fairly suppose, that the same proportion may hold through the whole extent of the universe.

It is clear that with this immense void at their disposal: [37]

. . . every single particle would have a sphere of void space around it 8,575 hundred thousand million million times bigger than the dimension of that particle.

Accordingly, Democritian atoms, whatever their initial disposition in space, would pretty soon be completely dispersed and would be unable to form even the most simple bodies, and much less, of course, such an artful and well-ordered system as, for instance, our solar world. Fortunately for its — and for our — existence, atoms are not free and independent of each other but are bound together by mutual gravitation.

Now this is already a refutation of atheism — Bentley, as we have seen, has learnt from Newton that gravitation cannot be attributed to matter — as it is clear [38]

> that such a mutual gravitation or spontaneous attraction can neither be inherent and essential to matter, nor ever supervene to it, unless impressed and infused into it by a divine power,

just because action at a distance [39]

> . . . is repugnant to common sense and reason. 'Tis utterly inconceivable, that inanimate brute matter, without the mediation of some immaterial being, should operate upon and affect other matter without mutual contact; that distant bodies should act upon each other through a *vacuum*, without the intervention of something else, by and through which the action may be conveyed from one to the other. We will not obscure and perplex with multitude of words what is so clear and evident by its own light, and must needs be allowed by all that have competent use of thinking, and are initiated into, I do not say the mysteries, but the plainest principles of philosophy. Now, mutual gravitation or attraction, in our present acceptation of the words, is the same thing with this, 'tis an operation, or virtue, or influence of distant bodies upon each other through an empty interval, without any *effluvia*, or exhalations, or other corporeal medium to convey and transmit it. This power, therefore, cannot be innate and essential to matter: and if it be not essential, it is consequently most manifest, since it doth not depend upon motion or rest, or figure or position of parts, which are all the ways that matter can diversify itself, that it could never supervene to it, unless impressed and infused into it by an immaterial and divine power.

Now, if we admit, as we must do, that this mutual

attraction cannot be explained by any "material and mechanical agent," the indubitable reality of this power of mutual gravitation [40]

> . . . would be a new and invincible argument for the being of God, being a direct and positive proof that an immaterial living mind doth inform and actuate the dead matter and support the frame of the world.

Moreover, even if reciprocal attraction were essential to matter, or if it were simply a blind law of action of some immaterial agent, it would not suffice to explain the actual fabric of our world, or even the existence of any world whatever. Indeed, under the unhampered influence of mutual gravitation, would not all matter convene together into the middle of the world?

Bentley seems to have been rather proud of having found that God not only pulled or pushed bodies towards each other, but also counteracted His action — or, more simply, suspended it — in the case of the fixed stars, at least of the outermost ones, which He prevented in this manner from leaving their places and maintained at rest.

Alas, Newton explained to him that his reasoning implied a finite world and that there was no reason to deny its possible infinity, that the difficulties Bentley found in the concept of an infinite sum or series were not contradictions, and that his refutation of the infinity (or eternity) of the world was a paralogism. Newton confirmed, however, that even in the case of an infinite world the mere and pure action of gravity could not explain its structure, and that choice and purpose were clearly apparent in the actual distribution of the heavenly bodies in space, as well as in the mutual adjustment of their masses, velocities and so on: [41]

As to your first query, it seems to me that if the matter of our sun and planets, and all the matter of the universe, were evenly scattered throughout all the heavens, and every particle had an innate gravity towards all the rest, and the whole space throughout which this matter was scattered was but finite; the matter on the outside of this space would, by its gravity, tend towards all the matter on the inside, and, by consequence, fall down into the middle of the whole space, and there compose one great spherical mass. But if the matter was evenly disposed throughout an infinite space, it could never convene into one mass; but some of it would convene into one mass, and some into another, so as to make an infinite number of great masses, scattered at great distances from one to another throughout all that infinite space. And thus might the sun and fixed stars be formed, supposing the matter were of a lucid nature. But how the matter should divide itself into two sorts, and that part of it which is fit to compose a shining body should fall down into one mass and make a sun, and the rest which is fit to compose an opaque body should coalesce, not into one great body, like the shining matter, but into many little ones; or if the sun at first were an opaque body like the planets, or the planets lucid bodies like the sun, how he alone should be changed into a shining body, whilst all they continue opaque, or all they be changed into opaque ones, whilst he remains unchanged; I do not think explicable by mere natural causes, but am forced to ascribe it to the counsel and contrivance of a voluntary Agent.

. . . .

To your second query, I answer, that the motions which the planets now have could not spring from any natural cause alone, but were impressed by an intelligent Agent. For since comets descend into the region of our planets, and here move all manner of ways, going sometimes the same way with the planets, sometimes the contrary way, and sometimes in cross ways, in planes inclined to the plane of the ecliptic, and at all kinds of angles, 'tis plain that there is no natural cause which could determine all the planets, both primary and secondary, to move the same way and in the same plane, without any considerable variation: this must have been the effect of counsel. Nor is there any natural cause which could give the planets those just degrees of velocity, in proportion to their distances from the sun and other central bodies, which were requisite to make them move in such concentric orbs about those bodies.

. . . .

To make this system, therefore, with all its motions, required a cause which understood and compared together the quantities of matter in the several bodies of the sun and planets, and the gravitating powers resulting from thence; the several distances of the primary planets from the sun, and of the secondary ones from Saturn, Jupiter, and the earth; and the velocities with which these planets could revolve about those quantities of matter in the central bodies; and to compare and adjust all these things together, in so great a variety of bodies, argues that cause to be, not blind and fortuitous, but very well skilled in mechanics and geometry.

Having learnt his lesson, Bentley writes therefore: [42]

. . . we affirm, that though we should allow that reciprocal attraction is essential to matter, yet the atoms of a chaos could never so convene by it as to form the present system; or, if they could form it, yet it could neither acquire these revolutions, nor subsist in the present condition, without the conservation and providence of a divine Being.

I. For, first, if the matter of the universe, and consequently the space through which it's diffused, be supposed to be *finite*, (and I think it might be demonstrated to be so, but that we have already exceeded the just measures of a sermon,) then, since every single particle hath an innate gravitation toward all others, proportioned by matter and distance; it evidently appears, that the outward atoms of the chaos would necessarily tend inwards, and descend from all quarters toward the middle of the whole space. For, in respect to every atom, there would lie through the middle the greatest quantity of matter and the most vigorous attraction; and those atoms would there form and constitute one huge spherical mass, which would be the only body in the universe. It is plain, therefore, that upon this supposition the matter of the chaos could never compose such divided and different masses as the stars and planets of the present world.

Furthermore, even if the matter of the chaos could build the separate bodies of the planets, they " could not possibly acquire such revolutions in circular orbs, or in ellipses very little eccentric," as they actually perform, by the mere action of the forces of inertia and gravity, and finally, " if we should grant . . . that these circular revolutions could be naturally attained," it still requires a divine power and providence to preserve them, and, generally speaking, to preserve the fabric of the world. For, even if we admitted that the combination of inertia and gravity would suffice for the maintaining of the orbital motion of the planets, what about the fixed stars? What prevents them from coming together? " If the fixed stars . . . are supposed to have no power of gravitation, 'tis plain proof of divine Being " as it shows the non-natural character of gravitation. "And 'tis as plain a proof of a divine Being if they have the power of gravitation." For, in that case, only a divine power can compel them to remain in their assigned places. But what if the world were not finite, but infinite? According to Bentley it does not make a very great difference: [43]

> . . . in the supposition of an *infinite* chaos, 'tis hard indeed to determine what would follow in this imaginary case from an innate principle of gravity. But, to hasten to a conclusion, we will grant for the present, that the diffused matter might convene into an infinite number of great masses, at great distances from one another, like the stars and planets of this visible part of the world. But then it is impossible that the planets should naturally attain these circular revolutions, either by principle of gravitation, or by impulse of ambient bodies. It is plain there is no difference as to this, whether the world be infinite or finite; so that the same arguments that we have used before may be equally urged in this supposition.

In spite of these clear proofs of God's purposeful action in the world, there are, as we know, people who refuse to be convinced by them and who argue that an infinite world can have no purpose. What indeed can be the usefulness of these innumerable stars that are not even seen by us, either by the unassisted eye or through the strongest telescope? But, replies Bentley, embracing the pattern of reasoning based on the principle of plenitude, "We must not confine and determine the purposes in creating all mundane bodies merely to human ends and uses." For, though, as it is evident, they are not created for our sakes, they are certainly not made for their own: [44]

> For matter hath no life nor perception, is not conscious of its own existence, not capable of happiness, nor gives the

> sacrifice of praise and worship to the Author of its being. It remains, therefore, that all bodies were formed for the sake of intelligent minds: and as the earth was principally designed for the being and service and contemplation of men, why may not all other planets be created for the like uses, each for their own inhabitants which have life and understanding? If any man will indulge himself in this speculation, he need not quarrel with revealed religion upon such account. The holy Scriptures do not forbid him to suppose as great a multitude of systems, and as much inhabited as he pleases. . . . God Almighty, by the inhausted fecundity of his creative power, may have made innumerable orders and classes of rational minds; some in their natural perfections higher than human souls, others inferior.

An indefinitely extended and populated world, immersed in an infinite space, a world governed by the wisdom and moved by the power of an Almighty and Omnipresent God, such is, finally, the universe of the very orthodox Richard Bentley, future Bishop of Worcester and Master of Trinity College. Such is, doubtlessly too, the universe of the very heretical Lucasian Professor of Mathematics, Isaac Newton, Fellow of the Royal Society and of the same Trinity College.[45]

VIII. The Divinization of Space

Joseph Raphson

Newton, as far as I know, never quoted More; nor did he make an explicit reference to his teachings. Yet the relations between the theories of the two Cambridge men could not, of course, escape their contemporaries. It is therefore not surprising that, fifteen years after the publication of the *Mathematical Principles of Natural Philosophy*, their connection was openly proclaimed by Joseph Raphson, a promising young mathematician, Master of Arts and Fellow of the Royal Society,[1] in an extremely interesting and valuable Appendix which he added, in 1702, to the second edition of his *Universal Analysis of Equations*.[2]

In this *Appendix*, which bears the title *On the real space or the Infinite Being*, Joseph Raphson, who obviously has neither Newton's subjective inclination for reticence and secrecy, nor his objective reasons for prudence, dots all the i's and crosses all the t's.

Starting with a historical account of the development of the conception of space which begins with Lucretius and culminates in Henry More's criticism of the Cartesian identification of extension with matter, his characterization of matter by impenetrability, and his demonstration of the existence of an immovable and immaterial extension, Raphson states his conclusion: [3]

> Thus from every motion (extended and corporeal), even from the [only] possible ones follows necessarily [the existence of] an immovable and incorporeal extended [entity], because everything which moves in the extension must necessarily move through extension. The extension of the real motion demonstrates the real existence of this immovable extended [entity], because otherwise it [the motion] can be neither expressed nor conceived, and because that which we cannot but conceive is necessarily true. It could be argued in the same manner concerning the supposed motion of figures in geometry. The possibility of these motions demonstrates the hypothetical necessity of this immovable extended [entity], and the reality of the physical motions, the absolute.

There is an unmistakable Spinozistic flavor in Raphson's terminology and manner of speaking. Yet, though deeply influenced by Spinoza,[4] Raphson is by no means Spinozist. On the contrary, More's distinction between the infinite, immovable, immaterial extension and the material, mobile and therefore finite one is, according to him, the sole and only means of avoiding the Spinozistic identification of God with the world. But let us proceed with Raphson's presentation of Henry More's theories.

The existence of motion implies, indeed, not only the distinction between the immovable, immaterial extension and the material one, and thus the rejection of the Cartesian identification; it implies also the rejection of the Cartesian negation of *vacuum*: in a world completely and continuously filled with matter rectilinear motion would be utterly impossible, and even circular motion would be extremely difficult to achieve.[5] The real existence of really void spaces can thus be considered as fully demonstrated. Wherefrom we can draw the following corollaries: [6]

> 1. The universal mass of movable [bodies] (or of the world) must necessarily be finite, because, on account of the vacuum and the mobility, each and every system of it may be compressed into a smaller place; the finitude of the ensemble of these systems, that is, of the world, follows herefrom necessarily, though the human mind will never be able to arrive at its limit.
>
> 2. All the finite [beings] existing separately can be comprehended by a number. It is possible that no created mind is able to comprehend it. Nevertheless, to their numerating Author, they will be in a finite number: this can also be shown as follows: let, for example, (*a*) be the minimum of what can exist, then (*a*) infinitely multiplied will turn out to be infinite; indeed, if it gave a finite sum

the true minimum (or atom) would not be (*a*) but another infinitely smaller, or infinitely small, body. This, however, as Raphson states, is "against the hypothesis." Of course we are not studying here the composition of space: we are dealing with impenetrable extended beings, that is, with bodies.

> 3. Herefrom can be argued the falsity of the teaching of Spinoza, who, misusing his 6th definition, makes it so wide as to force matter, insofar as it expresses essence, to express the essence of the Infinite Being, and to be one of its attributes. I recognize, however, and I can demonstrate, that everything of which the essence implies an absolute

infinity pertains necessarily to the *absolutely Infinite* Being; it is in this way that I derive my idea of the absolutely Infinite Being, which involves the supreme and absolute necessity.

The error of Spinoza is thus at once elucidated and corrected. Raphson obviously thinks that Spinoza was perfectly right in following the (Cartesian) principle of attributing to God all that is essentially infinite; right also in rejecting the Cartesian distinction between the infinite and the indefinite and in claiming for His extension actual and not only potential infinity. But he is wrong in accepting the Cartesian identification of extension and matter. Owing to Henry More's criticism of Descartes, Raphson believes he is able to escape the Spinozistic conclusion by attributing the infinite, *immaterial* extension to God, and reducing matter to the status of creature.

Matter, as we know, is characterized by Raphson by its mobility (which implies finitude) and impenetrability. As for the immaterial extension, or more simply, space, its properties, nature and existence are derived by him *more geometrico* "from the necessary and natural concatenation of simple ideas."[7]

Space is defined as[8] "the innermost extended [entity] (whatever it be) which is the first by nature and the very last to be obtained by continuous division and separation"; Raphson informs us that it is an imperfect definition or description of the defined object; it does not tell us anything about its essence, but, on the other hand, it has the advantage of being immediately acceptable as designating something the existence of which is perfectly evident and indubitable. Moreover, the analysis of the ideas used in this definition will lead us towards important consequences, namely towards the affirmation of the existence of a *real space* really distinct from matter.

The investigation starts with a postulate, according to which a " given idea " always enables us to derive from it the properties of the object, even making abstraction of its existence. Three corollaries are added, and these tell us that: [9]

> *All finite extended* can be divided (if only by the mind) or, what is the same, be conceived as divided.
>
> And it is (even if only for the concept) movable and possesses an actual figure.
>
> And [its] parts can be separated or removed from each other (if only by the mind), or be conceived as being removed.

An axiom then asserts that: [10]

> Between things separated or removed from each other there is always a distance (be it great or small), that is *something extended.*

A series of propositions now follows in quick succession: [11]

> 1. *Space* (or the *innermost extended*) is by its nature, and absolutely, indivisible, nor can it be conceived as divided

— which, if division means separation and mutual removal of parts, that is, divisibility means *discerpibility,* is, of course, a cogent consequence of the above-quoted corollaries.

> 2. *Space* is absolutely, and by its nature, immovable

— motion indeed implies divisibility.

3. *Space* is actually infinite

— which; *vice versa*, implies, immediately and by necessity, its absolute immovability.

4. *Space* is pure act.
5. *Space* is all-containing and all-penetrating.

To pave the way to further development, that is, to the identification of space with an attribute of God, Raphson adds that[12]

> . . . doubtlessly this is the reason why for the Hebrews the name of this *Infinite* was *Makom*; as it is that of St Paul's 'it is nearer to us than we are to ourselves.' It is to this Infinite that assuredly must be referred a great number of passages of the Holy Scripture as well as the hidden wisdom of the old Hebrews about the highest and incomprehensible amplitude of the *Ensoph*; as well as the teaching of the Gentiles about the all permeant, the all-embracing etc.

But let us not think that space is a kind of immaterial stuff — Raphson, obviously, wants to oppose *space* to More's *spirit*: [13]

> It is patent that space is not penetrated by anything: being infinite and undivided it penetrates everything by its innermost essence, and therefore cannot itself be penetrated by anything, nor even can it be conceived as penetrated.

It is clear thus that[14]

6. *Space* is incorporeal.
7. *Space* is immutable.

8. *Space* is one in itself, [and therefore] . . . it is the most simple entity, not composed of any things and not divisible into any things.

9. *Space* is eternal [because] the actually infinite cannot not be . . . in other words, that *it cannot not be* is essential to the actually infinite. It was therefore always.

This means that it is, or has, a necessary being, that the eternity of the infinite is the same thing as its existence, and that both imply the same necessity.[15]

10. *Space* is incomprehensible to us, [just because it is infinite].

11. *Space* is most perfect in its kind [*genus*].

12. Extended things can neither be nor be conceived without it. And therefore

13. *Space* is an attribute (namely the immensity) of the First Cause.

This last proposition, according to Raphson, can also be demonstrated in a much easier and more direct way: as, indeed, the First Cause[16]

. . . can neither give anything that it does not possess, nor be the cause of any perfection that it does not contain (in a certain manner) in the same degree if not in a greater one; and as there can be nothing *in rerum natura* except extended and unextended [things]; and as we have demonstrated that extension is perfection, existing everywhere, and is even infinite, necessary, eternal, etc., it follows necessarily that it must be found in the *First Cause* of the extended [things] without which the extended [things] cannot exist. Which it was proper to demonstrate. For the true and reciprocal reason of the omniform, true and actual infinity is found to consist in the most absolute unity, just as,

> *vice versa*, the highest reason of the unity culminates in and is absorbed by the infinity. For whatever expresses the actual, and in its kind most absolute, infinity, necessarily expresses the essence of the *First Cause*, the Author of everything that is.

It is rather curious to see Raphson use the Cartesian and even Spinozistic logic and patterns of reasoning to promote Henry More's metaphysical doctrine. Yet it cannot be denied that by these means Raphson succeeded in giving it a much higher degree of consistency than it had from its author. Henry More, indeed, could only present us with a list of " titles " applicable both to space and to God. Raphson shows their inner connection; moreover, by identifying infinity, on the one hand, with highest perfection, and, on the other hand, by transforming extension itself into perfection, he makes the attribution of extension to God logically as well as metaphysically unavoidable.

Having established the attribution to the *First Cause* of infinite space (which taken abstractly is the object of geometry, and taken as reality is the very immensity of God), Raphson now goes on to a more careful consideration of their connection: [17]

> That its [the First Cause's] true and essential presence is a necessary prerequisite as well of the essential being as of the real existence of all things is recognized by a number of contemporaries. But, how this essential and intimate presence can be explained in the hypothesis of the non-extension [of the First Cause] without a manifest contradiction has not yet been made clear; and it will never be possible to make it clear. Indeed, to be present by essence in places diverse and distant from each other, for instance

> in the globe of the Moon and in that of the earth, and also in the intermediate space, what else is it but, precisely, to extend oneself? Now, we have demonstrated that this extension is truly real, indivisible, immaterial (or, if you wish, spiritual). What else is there to be desired in order to infer its perfection, supreme and infinite of its kind (insofar as it is an inadequate concept of the Infinite Being)?

I do not see, concludes Raphson, by what other name than extension or space this essential omnipresence of the First Cause could be expressed.

The philosophers were right, of course, in removing from the First Cause the imperfect, divisible, material extension. Yet, by the rejection from it of all kinds of extension, they opened up the way towards atheism, or rather hylotheism, to a great many people, namely, to those who did not want to be hemmed in by ingenuous circuits of ambiguous circumlocutions and embarrassed by obscure and unintelligible notions and terms. Such are Hobbes and some others: because they did not find anywhere in the world this infinite and eternal, unextended Supreme Being, they thought that it did not exist at all, and boldly proposed their opinions to the world. So too had some of the ancients, who insisted upon the incomprehensibility of the Supreme Being. The explanation of all these aberrations is to be sought, according to Raphson, in the misunderstanding of the very essence of extension that has been falsely held to be necessarily something imperfect and lacking all unity and reality. In truth, however, extension, as such, is something positive and denotes a very real perfection. Accordingly, as generally [18]

> . . . everything positive and substantial that is found in the essence of things as their primary and constitutive attribute, such as extension in matter, etc., must necessarily be really and truly present in the First Cause, and be in it in a degree of infinite excellence in the manner most perfect of its kind,

the infinite extension must be truly and really, and not only metaphorically, attributed to the First Cause.

The First Cause appears thus as the twofold source, or cause, of the perfections of the created things that it contains, as the Schoolmen say, in an eminent and transcendent manner.[19]

> For (as they say) *it gives nothing that it does not have* (in a more perfect manner) *in itself.*

Consequently they assert that God is a thinking Being: how could, indeed, *a thinking being* (*like ourselves*) *proceed from a non-thinking one?* But we can reverse the question and, with exactly the same right, ask: *how can an extended being come forth from an unextended one?* The Schoolmen, of course, want both perfections to be contained in the First Cause in the transcendent manner. As for extension, such as it is in matter, they justly argue that it is imperfect. We, however, and we can quote good authorities in favor of this opinion, for instance, Father Malebranche, regard cogitation, or thought (such as it is in human minds, or in the created spirits), to be just as imperfect in comparison to that of the Absolutely Infinite Being. And though, perhaps, cogitation in finite thinking beings is much more perfect than extension, as it is in matter, it is doubtless removed by the same interval, that is, by an infinite one, from the source of these perfections in the First Cause, and, in relation to it, they are both equally imperfect.[20]

> The infinite amplitude of extension expresses the immense diffusion of being in the First Cause, or its infinite and truly interminate essence. This [amplitude] is that originary *extensive* perfection, which we have found, so imperfectly counterfeited, in matter.
>
> The infinite (whatever it be) and most perfect energy, everywhere indivisibly the same, which produces and perpetually conserves everything (and which this never-sufficiently-to-be-admired series of *Divine Ratiocination*, that is, the whole fabric of nature, more than sufficiently demonstrates to us *a posteriori*), is this *intensive* perfection, which though [distant from it] by an infinite interval in kind as well as in degree, we, miserable examples of the infinite Archetype, flatter ourselves to imitate.

Raphson's assertions are to be taken *verbatim*: extension as such is a perfection, even gross, material extension. The *modus* of its realization in bodies is, to be sure, extremely defective, precisely as our discoursive thought is an extremely defective *modus* of cogitation; but, just as in spite of its discoursiveness our thought is an imitation of, and a participation in, God's cogitativeness, so in spite of its divisibility and mobility our bodily extension is an imitation of, and a participation in, God's own and perfect extensiveness.

As for the latter, we have already proved that: [21]

> . . . this internal or truly innermost *locus* penetrates everything by its essence and, undivided, is most intimately present in everything; that it cannot be, or even be thought of, as penetrated by any thing, and that it is infinite, most

perfect, one and indivisible. Hence it clearly appears by what infinite interval are distant from it all other things that have only an evanescent being and, to use the elegant expression of the Prophet (Isaiah, 40), are as nothing to this *Infinite* and *Eternal* and, so to speak, essential (οὐσιότατον) Being. They are, as it were, light shadows of the true Reality and even if they were everywhere, they would by no means express even in the lowest degree that Infinity which we understand to be supremely positive and supremely real in the First Cause.

Thus, even if it were infinitely extended — which it is not — matter would never be identical with the divine extension and would never be able to become an attribute of God. Joseph Raphson is to such a degree elated and ravished by the contemplation of the idea of infinity that we could apply to him (though modifying it somewhat) the expression used by Moses Mendelssohn for Spinoza: he is drunk with infinity. He goes so far as — paradoxically — to reject Henry More's reassertion of the fundamental and primary validity of the category or question: " where? " In infinity it has no meaning. The infinite is not something, a sphere, of which the center is everywhere and the limits nowhere; it is something of which the center is nowhere also, something that has neither limits nor center, something in respect to which the question " where? " cannot be asked, as in respect to it everywhere is nowhere, *nullibi.*[22]

In respect to this immense *locus* a system of finite bodies, be it ever so large, is truly said to be nowhere. It is indeed utterly immeasurable; *here, there, in the middle,* etc. vanish in it completely.

Raphson is obviously right. In the infinite homogeneous space all "places" are perfectly equivalent and cannot be distinguished from each other: they all have the same "position" in respect to the whole.[23]

> The illustrious Guericke has very well written about it in his *Magdeburgian Experiments*, p. 65: If in this immensity (which has no beginning, nor end, nor middle) somebody marched for an infinitely long [time], and traversed innumerable thousands of miles, he would, in relation to this immensity, be in the same place; and if he repeated his action and arrived ten infinities farther, he would nevertheless be in this immensity in the same way and in the same place and would not be a single step nearer to the end, or the fulfillment of his intention, because in the Immeasurable (*Immensum*) there is no relation. All relations in it are conceived in reference to ourselves or to some other created thing. Indeed this immense *locus* is truly everywhere; and everything that has its finite *where?* (as they are wont to speak about spirits) has it as a relation to some other finite [thing]; but in relation to the Immensity it is truly nowhere.

Yet, if Raphson insists so strongly upon the infinity of uncreated space in contradistinction to the finitude of the created world, it is by no means his intention to assign to this latter determinate, or even determinable — by us — dimensions. Quite the contrary: in infinite space there is room enough for a practically indeterminate and indefinitely large world. Thus he tells us that if[24]

> . . . there can be absolutely no reason why [the world] should extend itself to the infinity of its immense *locus*, as it does not possess an absolute plenitude and is composed of movable parts . . . whereas the *absolutely Infinite*

> is utterly immovable and absolutely one or full of itself . . . [nevertheless] . . . how great the universe is or how far it extends, is completely hidden to us.

Raphson himself would[25]

> . . . easily believe that it can be immeasurable in respect to our capacity of understanding, and that we shall never be able to comprehend it. Indeed, it does not follow that we can comprehend by our cogitation all magnitude that is not infinite, or that we should ever be able to depict it in our mind as so large that the universe could not, in truth, be even larger. We can, for instance, conceive a series of numbers, disposed in a straight line, to extend from this our earth to the Dog-Star, or to any one in the Milky Way or to whatever visible limit, the unity of these [numbers] expressing the distance between the earth and that limit; we can also conceive this number to be squared, raised to the third, fourth, and so on, power, until the index of this power becomes equal to the first number, or to its first root; we can finally consider this power as a root of others, progressing in the same manner. And yet it is, perhaps, as nothing compared to the magnitude of the universe which can, and possibly does, surpass the capacity of any finite numbering [mind], not only ours, and cannot be comprehended by any other than its immense Author. Yet it is certain that it cannot be infinite in that absolute manner in which the *First Cause* is, insofar as it is considered as the immense *locus* of things.

We see it thus quite clearly: the difference between the infinite and the finite is not a difference between " more " and " less "; it is not a quantative, but a qualitative one, and, though studied by mathematicians, it is fundamentally a metaphysical difference. It is this difference which, fully understood, enables us not to lapse into the error of a pantheistic confusion of the Creator God with the created world, and it is this selfsame difference which provides us with a firm ground for the study of the nearly infinite variety of created things. Indeed, those [26]

> who will [study] this part of the universe, visible to us, not only in books, but who will diligently read and carefully contemplate [the book of Nature], using his own observations and the [analysis] of the constitution of the skies, will hardly fail to recognize not only that there can be a plurality of worlds, but that, in truth, there are a nearly infinite number of systems, various laws of motion, exhibiting various (nearly innumerable) phenomena and creatures.

Why, even on this earth there are so many and such varied creatures, endowed with so many different faculties, possibly even with some that are completely unknown to us. How many more could there be elsewhere that can be called into being by the infinite combinative art of the Infinite Architect.

As for us, the only doors open to the true cogitation of the universe are observation and experience. By the first we arrive at the system of visible motions of the world; by the second we discover the forces, the (sensible) qualities and mutual relations of bodies. Mathematics (mathematical physics) and chemistry are the sciences that arise on these empirical foundations. As for the "hypotheses" that go beyond these empirical data, they may be plausible, and even, sometimes, useful for the investigation of truth; yet they breed prejudices and therefore cause more harm than good. *Hypothesomania*, the invention of new hypotheses, belongs to poetical and fictitious philosophy, not to the pursuit of knowledge.

For the latter, according to Raphson, the method established by the supreme philosopher, Newton, in his *Principia*, consisting in the study of the phenomena of nature by means of experiments and rational mechanics, reducing them to forces the action of which — though their nature is hidden from us — is obvious and patent in the world.

As we see, empiricism and metaphysics, and even a very definite kind of metaphysics, the creationist, are closely linked together. What other means, indeed, but observation and experience can we possibly use for the study of a world freely created by an Infinite God? Raphson concludes therefore: [27]

> Neither can Human Philosophy theoretically compose the smallest mouse or the simplest plant, nor can human praxis build them, much less the whole universe. These are problems worthy of the Primordial Wisdom and Power which produces these things. As for us, they offer us only a progress *in aeternum* of our knowledge both of the things themselves and of the perpetually geometrizing God.

IX. God and the World:

SPACE, MATTER, ETHER AND SPIRIT

Isaac Newton

It is difficult to tell what the reasons were that determined Newton to enlarge, in the Latin edition (translation) of his *Opticks*, the number of *Queries* appended by him to the third book of his work, and to include among the additional ones two rather long and extremely important and interesting papers which, in contradistinction to the purely technical *Queries* of the first English edition, deal, not with optical, but with methodological, epistemological and metaphysical problems.[1]

The publication of Raphson's book could not have been the motive: the *De spatio reali* was published in 1702, the Latin translation of the *Opticks* in 1706; but the English edition appeared in 1704 and if Newton wanted to make his position clear in relation to Raphson's, he could, and should have done it in 1704. It is possible, in my opinion — though it is only a conjecture — that it was the publication of Dr. George Cheyne's *Philosoph-*

ical Principles of Natural Religion that gave Newton the incentive, usually lacking, to come into the open.[2]

Now, be this as it may, it is these *Queries* (which, curiously enough, seem to have been ignored by Berkeley) which build the subject of the famous polemics between Leibniz and Clarke. It is, indeed, in these *Queries* (21 and 22) that, in a much more precise and clear manner than anywhere else — the *General Scholium* of the second edition of the *Principia* not excluded — Newton states his conceptions about the purpose and aim of philosophy and develops, at the same time, his general world-view: an extremely interesting and fairly consistent system of " corpuscular philosophy " — already sketched in his letters to Bentley — asserting the fundamental unity of matter and light, and presenting the material components of the universe, that is, hard, indivisible particles, as constantly acted upon by quite a system of various *non-material* attractive and repulsive forces. Thus *Query* 20 (28 in the second edition) explains at length the physical (astronomical) inadmissibility of the *plenum* (a completely full space would oppose such a strong resistance to motion that it would be practically impossible and would have ceased long ago), as well as the physical (astronomical) admissibility of the celestial spaces' being filled with an extremely thin, rare and tenuous ether, of which the density can be made as small as we wish (is not our air " at the height of 70, 140, 210 miles 100,000, 100,000,000,000 or 100,000,000,000,000 times rarer, and so on " than on the earth?), which implies the granular structure of this ether, the existence of a *vacuum* and the rejection of a continuous medium, and concludes: [3]

And for rejecting such a Medium, we have the Authority of those oldest and most celebrated Philosophers of *Greece* and *Phoenicia*, who made a *Vacuum*, and Atoms, and the Gravity of Atoms, the first Principles of their Philosophy; tacitly attributing Gravity to some other Cause than dense Matter. Later Philosophers banish the Consideration of such a Cause out of natural Philosophy, feigning Hypotheses for explaining all things mechanically, and referring other Causes to Metaphysicks: Whereas the main Business of natural Philosophy is to argue from Phaenomena without feigning Hypotheses, and to deduce Causes from Effects, till we come to the very first Cause, which certainly is not mechanical; and not only to unfold the Mechanism of the World, but chiefly to resolve these and such like Questions. What is there in places almost empty of Matter, and whence is it that the Sun and Planets gravitate towards one another, without dense Matter between them? Whence is it that Nature doth nothing in vain; and whence arises all that Order and Beauty which we see in the World? To what end are Comets, and whence is it that Planets move all one and the same way in Orbs concentrick, while Comets move all manner of ways in Orbs very excentrick; and what hinders the fix'd Stars from falling upon one another? How came the Bodies of Animals to be contrived with so much Art, and for what ends were their several Parts? Was the Eye contrived without Skill in Opticks, and the Ear without Knowledge of Sounds? How do the Motions of the Body follow from the Will, and whence is the instinct in Animals? Is not the Sensory of Animals that place to which the sensitive Substance is present, and into which the sensible Species of Things are carried through the Nerves and Brain, that there they may be perceived by their immediate presence to that Substance? And these things being rightly dispatch'd, does it not appear from Phaenomena that there is a Being incorporeal, living, intelligent, omnipresent, who in infinite Space, as it were in his Sensory, sees the things themselves intimately, and thoroughly perceives them, and comprehends them wholly by their immediate presence to himself: Of which things the Images only carried through the Organs of Sense into our little Sensoriums, are there seen and beheld by that which in us perceives and thinks. And though every true Step made in this Philosophy brings us not immediately to the Knowledge of the first Cause, yet it brings us nearer to it, and on that account is to be highly valued.

As for *Query* 23 (31), it starts with the question: [4]

Have not the small Particles of Bodies certain Powers, Virtues, or Forces, by which they act at a distance, not only upon the Rays of Light for reflecting, refracting, and inflecting them, but also upon one another for producing a great Part of the Phaenomena of Nature? For it's well known, that Bodies act one upon another by the Attractions of Gravity, Magnetism, and Electricity; and these Instances shew the Tenor and the Course of Nature, and make it not improbable but that there may be more attractive Powers than these. For Nature is very consonant and conformable to her self.

Newton does not tell us outright — any more than he does in the *Principia* — what these various "Powers" are. Just as in the *Principia*, he leaves that question open, though, as we know, he holds them to be non-mechanical, immaterial and even "spiritual" energy extraneous to matter.[5]

How these Attractions may be perform'd, I do not here consider. What I call attraction may be perform'd by impulse, or by some other means unknown to me. I use that Word here to signify only in general any Force by which Bodies tend towards one another, whatsover be the Cause. For we must learn from the Phaenomena of Nature what Bodies attract one another, and what are the Laws and Properties of the Attraction, before we enquire the Cause by which the Attraction is perform'd. The Attractions of Gravity, Magnetism, and Electricity, reach to very sensible distances, and so have been observed by vulgar Eyes, and there may be others which reach to so small distances as hitherto escape Observation; and perhaps electrical Attraction may reach to such small distances, even without being excited by Friction.

Whatever these "Powers" may be, they are, in any case, real forces and perfectly indispensable for the explanation — even a hypothetical one — of the existence of bodies, that is, of the sticking together of the material particles that compose them; a purely materialistic pattern of nature is utterly impossible (and a purely materialistic or mechanistic physics, such as that of Lucretius or of Descartes, is impossible, too): [6]

> The Parts of all homogeneal hard Bodies which fully touch one another, stick together very strongly. And for explaining how this may be, some have invented hooked Atoms, which is begging the Question; and others tell us that Bodies are glued together by rest, that is, by an occult Quality, or rather by nothing; and others, that they stick together by conspiring Motions, that is, by relative rest amongst themselves. I had rather infer from their Cohesion, that their Particles attract one another by some Force, which in immediate Contact is exceeding strong, at small distances performs the chymical Operations above-mention'd, and reaches not far from the Particles with any sensible Effect.

It could be argued, of course (and was to be argued by Leibniz) that Newton is wrong to stick to the classical atomic conception of hard, impenetrable, indivisible last components of matter, a conception which implies great difficulties for dynamics. It is indeed, impossible to say what would happen if two absolutely hard bodies should collide. Let us take, for instance, two perfectly similar and perfectly hard, that is, absolutely unyielding and indeformable, bodies, and let them approach each other — the classical case of dynamics — with the same speed. What will they do after the impact? Rebound, as elastic bodies would do? Or stop each other as would be the case with inelastic ones? As a matter of fact, they should not do either — yet, *tertium non datur*. As we know, Descartes, in order to preserve the principle of conservation of energy, asserted the rebounding. But he was obviously wrong. If we admit, however, that they would stop each other, that is, that motion is lost in every impact, would not the world-machine run down very quickly and very quickly come to a stop? Should we not, in order to avoid these difficulties, discard completely the atomic conception and admit, for instance, that matter is infinitely divisible or that its " last " components are not hard atoms but soft, or elastic, particles, or even " physical monads "? Newton, therefore, continues [7]

> All bodies seem to be composed of hard Particles: for otherwise Fluids would not congeal; as Water, Oils, Vinegar, and Spirit or Oil of Vitriol do by freezing; Mercury by fumes of Lead; Spirit of Nitre and Mercury, by dissolving the Mercury and evaporating the Flegm; Spirit of Wine and Spirit of Urine, by deflegming and mixing them; and Spirit of Urine and Spirit of Salt, by subliming them

together to make Sal-amoniac. Even the Rays of Light seem to be hard Bodies; for otherwise they would not retain different Properties in their different Sides. And therefore Hardness may be reckon'd the Property of all uncompounded Matter. At least, this seems to be as evident as the universal Impenetrability of Matter. For all Bodies, so far as Experience reaches, are either hard, or may be harden'd; and we have no other Evidence of universal Impenetrability, besides a large Experience without an experimental Exception. Now if compound Bodies are so very hard as we find some of them to be, and yet are very porous, and consist of Parts which are only laid together; the simple Particles which are void of Pores, and were never yet divided, must be much harder. For such hard Particles being heaped together, can scarce touch one another in more than a few Points, and therefore must be separable by much less Force than is requisite to break a solid Particle, whose Parts touch in all the Space between them, without any Pores or Interstices to weaken their Cohesion. And how such very hard Particles which are only laid together, hold and that so firmly as they do, without the assistance of something which causes them to be attracted or press'd towards one another, is very difficult to conceive.

This "something," as we know, and as it is clear from the very texts I am quoting, cannot be other, smaller, "ethereal" particles, at least not in the last analysis, because the same question, that is, the question about their interaction, can obviously be raised concerning the "ethereal" particles themselves, and cannot be answered by postulating an ultra-ether, which moreover, would imply the existence of an ultra-ultra-ether, and so on.

Forces of attraction, and also of repulsion are therefore fundamental, though non-material, elements of nature: [8]

> There are therefore Agents in Nature able to make the Particles of Bodies stick together by very strong Attractions. And it is the Business of experimental Philosophy to find them out.

Thus we see it once more: good, empirical and experimental natural philosophy does not exclude from the fabric of the world and the furniture of heaven immaterial or transmaterial forces. It only renounces the discussion of their nature, and, dealing with them simply as causes of the observable effects, treats them — being a *mathematical* natural philosophy — as *mathematical* causes or forces, that is, as mathematical concepts or relations. It is, on the contrary, the *a priori* philosophy of the classical Greek atomists, who at least recognized the existence of void space and probably even the non-mechanical character of gravity, and of course that of Descartes, that is guilty of this exclusion and of the impossible attempts to explain everything by matter and motion. As for Newton himself, he is so deeply convinced of the reality of these immaterial, and, in this sense, transphysical forces, that this conviction enables him to devise a most extraordinary and truly prophetic picture of the general structure of material beings: [9]

> Now the smallest Particles of Matter may cohere by the strongest Attractions, and compose bigger Particles of weaker Virtue; and many of these may cohere and compose bigger Particles whose Virtue is still weaker, and so on for divers Successions, until the Progression end in the biggest Particles on which the Operations in Chymistry,

and the Colours of natural Bodies depend, and which by cohering compose Bodies of a sensible Magnitude. If the Body is compact, and bends or yields inward to Pression without any sliding of its Parts, it is hard and elastick, returning to its Figure with a Force rising from the mutual Attraction of its Parts. If the Parts slide upon one another, the Body is malleable or soft. If they slip easily, and are of a fit Size to be agitated by Heat, and the Heat is big enough to keep them in Agitation, the Body is fluid; and if it be apt to stick to things, it is humid; and the Drops of every fluid affect a round Figure by the mutual Attraction of their Parts, as the Globe of the Earth and Sea affects a round Figure by the mutual Attraction of its Parts by Gravity.

Moreover, as I have already hinted before, the admission of various immaterial forces acting upon or distributed around the bodies or particles according to strict mathematical laws — or to express it in a more modern way: the admission of the existence of different fields of forces connected with bodies and particles — enables us, and that is an invaluable advantage, to superimpose them one upon the other, and even to transform them into their contraries. Indeed,[10]

Since Metals dissolved in Acids attract but a small quantity of the Acid, their attractive Force can reach but to a small distance from them. And as in Algebra, where affirmative Quantities vanish and cease, there negative ones begin; so in Mechanicks, where Attraction ceases, there a repulsive Virtue ought to succeed. And that there is such a Virtue, seems to follow from the Reflexions and Inflexions of the Rays of Light. For the Rays are repelled by Bodies in both these Cases, without the immediate Contact of the reflecting or inflecting Body. It seems also to follow from the Emission of Light; the Ray so soon as it is shaken off from a shining Body by the vibrating Motion of the Parts of the Body, and gets beyond the reach of Attraction, being driven away with exceeding great Velocity. For that Force which is sufficient to turn it back in Reflexion, may be sufficient to emit it. It seems also to follow from the Production of Air and Vapour. The Particles when they are shaken off from Bodies by Heat or Fermentation, so soon as they are beyond the reach of the Attraction of the Body, receding from it, and also from one another with great Strength, and keeping at a distance, so as sometimes to take up above a Million of Times more space than they did before in the form of a dense Body. Which vast Contraction and Expansion seems unintelligible, by feigning the Particles of Air to be springy and ramous, or rolled up like Hoops, or by any other means than a repulsive Power.

Thus, the admission of immaterial " virtues " offers us an immediate and elegant solution of the most important and crucial problem of elasticity, or " springiness " of bodies; and *vice versa,* this very solution demonstrates the impossibility of explaining this property of bodies by purely mechanical means (as Descartes and Boyle tried to do) and therefore confirms the insufficiency of pure materialism not only for philosophy in general, but also for *natural* philosophy. As a matter of fact, without the immaterial Powers and Virtues, there would not be any Nature to philosophize about, because there would be no cohesion, no unity and no motion; or if there were, at the beginning, it would have ceased long ago. On the contrary, if we admit the double, material as well as immaterial, structure of Nature,[11]

. . . Nature will be very conformable to her self and very simple, performing all the great Motions of the heavenly Bodies by the Attraction of Gravity which intercedes those Bodies, and almost all the small ones of their Particles by some other attractive and repelling Powers which intercede the Particles. The *vis inertiae* is a passive Principle by which Bodies persist in their Motion or Rest, receive Motion in proportion to the Force impressing it, and resist as much as they are resisted. By this Principle alone there never could have been any Motion in the World. Some other Principle was necessary for putting Bodies into Motion; and now they are in Motion, some other Principle is necessary for conserving the Motion. For from the various Composition of two Motions, 'tis very certain that there is not always the same quantity of Motion in the World. For if two Globes joined by a slender Rod, revolve about their common Center of Gravity with an uniform Motion, while that Center moves on uniformly in a right Line drawn in the Plane of their circular Motion; the Sum of the Motions of the two Globes, as often as the Globes are in the right Line described by their common Center of Gravity, will be bigger than the Sum of their Motions, when they are in a Line perpendicular to that right Line. By this Instance it appears that Motion may be got or lost.[11a] But by reason of the Tenacity of Fluids, and Attrition of their Parts, and the Weakness of Elasticity in Solids, Motion is much more apt to be lost than got, and is always upon the Decay. For Bodies which are either absolutely hard, or so soft as to be void of Elasticity, will not rebound from one another. Impenetrability makes them only stop. If two equal Bodies meet directly *in vacuo*, they will by the Laws of Motion stop where they meet, and lose all their Motion, and remain in rest, unless they be elastick, and receive new Motion from their Spring.

Yet, even if they be elastic, they cannot be absolutely elastic, and thus, by each and every impact, some motion (that is, momentum) will be lost. And if the world were full, as the Cartesians want it to be, then the "vortical" motion assumed by Descartes would cease very quickly, because [12]

> . . . unless the Matter were void of all Tenacity and Attrition of Parts, and Communication of Motion (which is not to be supposed,) the Motion would constantly decay. Seeing therefore the variety of Motion that we find in the World is always decreasing, there is a necessity of conserving it and recruiting it by active Principles,

that is, in the last analysis by the constant action in the world of the Omnipresent and All-powerful God. Newton therefore continues: [13]

> All these things being consider'd, it seems probable to me, that God in the Beginning form'd Matter in solid, massy, hard, impenetrable, moveable Particles, of such Sizes and Figures, and with such other Properties, and in such Proportion to Space, as most conducted to the End for which he form'd them; and that these primitive Particles being Solids, are incomparably harder than any porous Bodies compounded of them; even so very hard, as never to wear or break in pieces; no ordinary Power being able to divide what God himself made one in the first Creation. While the Particles continue entire, they may compose Bodies of one and the same Nature and Texture in all Ages: But should they wear away, or break in pieces, the Nature of Things depending on them, would be changed. Water and Earth, composed of old worn Particles and Fragments of Particles, would not be of the same Nature and Texture now, with Water and Earth composed of entire

> Particles in the Beginning. And therefore, that Nature may be lasting, the Changes of corporeal Things are to be placed only in the various Separations and new Associations and Motions of these permanent Particles; compound Bodies being apt to break, not in the midst of solid Particles, but where those Particles are laid together, and only touch in a few Points.
>
> It seems to me farther, that these Particles have not only a *Vis inertiae*, accompanied with such passive Laws of Motion as naturally result from that Force, but also that they are moved by certain active Principles. . . .

and it is the action of these principles, or, more exactly, the action of God by means of these principles that gives to the world its structure and order, and it is this structure and order that enables us to recognize that the world is an effect of choice, and not chance or necessity. Natural philosophy — at least the good one, that is, the Newtonian and not the Cartesian — thus transcends itself and leads us to God: [14]

> . . . by the help of these Principles, all material Things seem to have been composed of the hard and solid Particles above-mention'd, variously associated in the first Creation by the Counsel of an intelligent Agent. For it became him who created them to set them in order. And if he did so, it's unphilosophical to seek for any other Origin of the World, or to pretend that it might arise out of Chaos by the mere Laws of Nature; though being once form'd, it may continue by those Laws for many Ages. For while Comets move in very excentrick Orbs in all manner of Positions, blind Fate could never make all the Planets move one and the same way in Orbs concentrick, some inconsiderable Irregularities excepted, which may have risen

from the mutual Actions of Comets and Planets upon one another, and which will be apt to increase, till this System wants a Reformation. Such a wonderful Uniformity in the Planetary System must be allowed the Effect of Choice. And so must the Uniformity in the Bodies of Animals. . . .

All that, and much more besides,[15]

. . . can be the effect of nothing else than the Wisdom and Skill of a powerful ever-living Agent, who being in all Places, is more able by his Will to move the Bodies within his boundless uniform Sensorium, and thereby to form and reform the Parts of the Universe, than we are by our Will to move the parts of our own Bodies. And yet we are not to consider the World as the Body of God, or the several Parts thereof, as the Parts of God. He is an uniform Being, void of Organs, Members or Parts, and they are his Creatures subordinate to him, and subservient to his Will; and he is no more the Soul of them, than the Soul of Man is the Soul of the Species of Things carried through the Organs of Sense into the place of its Sensation, where it perceives them by means of its immediate Presence, without the Intervention of any third thing. The Organs of Sense are not for enabling the Soul to perceive the Species of Things in its Sensorium, but only for conveying them thither; and God has no need of such Organs, he being everywhere present in the Things themselves. And since Space is divisible *in infinitum*, and Matter is not necessarily in all places, it may be also allow'd that God is able to create Particles of Matter of several Sizes and Figures, and in several Proportions to Space, and perhaps of different Densities and Forces, and thereby to vary the Laws of Nature, and make Worlds of several sorts in several Parts of the Universe. At least, I see nothing of Contradiction in all this, concludes Newton, who could have added that in the *Principia* he had already shown — without insisting upon it — that the inverse square law of attraction, the actual law of this world, was by no means the only possible — although the most convenient one — and that God, had He wanted to, could have adopted another. As he could have quoted his friend Robert Boyle who believed that God had actually tried out, in different worlds, different laws of motion; or Joseph Raphson who had just expressed the same opinion. Yet he did not. As he did not quote Henry More when he made infinite space the *sensorium* of the nevertheless transcendent God.

X. Absolute Space and Absolute Time:

GOD'S FRAME OF ACTION

Berkeley

& Newton

It is certainly Raphson's interpretation, or, it would be better to say, Raphson's disclosure of the metaphysical background of Newtonianism, that Bishop Berkeley had in mind, when, in 1710, in his *Principles of Human Knowledge*, he not only made a vigorous attack upon its fundamental concepts, absolute space and absolute time, but also pointed out the great danger that they implied from the theological point of view. One of the chief advantages of the radical immaterialistic and sensualistic empiricism advocated by Berkeley is, according to him, the possibility it gives us of getting rid of these entities, asserted in [1]

> . . . a certain celebrated treatise of *mechanics*: in the entrance of which justly admired treatise, time, space and motion, are distinguished into *absolute* and *relative*, *true* and *apparent*, *mathematical* and *vulgar*; which distinction, as it is at large explained by the author, doth suppose those quantities to have an existence without the mind; and

that they are ordinarily conceived with relation to sensible things, to which nevertheless, in their own nature, they bear no relation at all.

"This celebrated author," continues Berkeley, who offers us a very precise account (largely in Newton's words) of the theory he is about to criticize, holds that [2]

> . . . there is an *absolute space*, which being unperceivable to sense, remains in itself similar and immovable, and relative space to be the measure thereof, which being movable, and defined by its situation in respect of sensible bodies, is vulgarly taken for immovable space.

Berkeley, of course, does not accept this theory; an unperceivable reality is unthinkable and "*philosophic considerations of motion doth not imply the being of absolute space* distinct from what is perceived by sense and related to bodies," Newton's assertions to the contrary notwithstanding. Moreover, and though last, not least,[3]

> What is here laid down seems to put an end to all those disputes and difficulties which have sprung up amongst the learned concerning the nature of *pure space*. But the chief advantage arising from it is, that we are freed from that dangerous *dilemma*, to which several who have employed their thoughts on this subject imagine themselves reduced, to wit, of thinking either that real space is God, or else that there is something beside God which is eternal, uncreated, infinite, indivisible, immutable. Both of which may justly be thought pernicious and absurd notions. It is certain that not a few divines, as well as philosophers of great note, have, from the difficulty they found in conceiving either limit or annihilation of space, concluded it must be

divine. And some of late have set themselves particularly to show, that the incommunicable attributes of God agree to it. Which doctrines, how unworthy soever it may seem of the divine nature, yet I do not see how we can get clear of it, so long as we adhere to the received opinions.

Berkeley's attack, though it certainly did not affect Newton as strongly as was thought by some of his historians, seems nevertheless to have been the reason, or at least one of the reasons — the second being Leibniz's accusation of introducing, by his theory of universal gravitation, the use of a senseless occult quality into natural philosophy [4] — that induced Newton to add to the second edition of his *Principia* the famous *General Scholium* which expresses so forcefully the religious conceptions that crown and support its empirico-mathematical construction and thus reveal the real meaning of his " philosophical " method. It seems to me rather probable that he wanted to dissociate himself from the somewhat compromising allies hinted at by Berkeley [4a] and, by exposing his views in his own manner, to demonstrate — as Bentley had already attempted to do — that natural philosophy, that is, *his* natural philosophy, leads necessarily not to the denial but to the affirmation of God's existence and of his action in the world. At the same time he obviously does not want to disavow or reject them, and in spite of Berkeley's warning, he asserts not only the existence of absolute time and space but also their necessary connection with God.

Compared to the statements made by Newton in his letters to Bentley — and much more so if compared to Bentley's elaboration of these statements and Newton's own developments in the *Queries* of the *Opticks* — Newton's pronouncements in the *General Scholium*, at least those concerning God's action in the world, are not very explicit. Thus, Newton does not tell us anything about the necessity of God's continuous concourse for the preservation of its structure; he seems even to admit that, once started, the motion of the heavenly bodies could continue forever; it is only at their beginning that God's direct intervention appears indispensable. On the other hand, the actual structure of the world (that is, of the solar system) is, of course, asserted to be the result of a conscious and intelligent choice: [5]

> . . . in the celestial spaces where there is no air to resist their motions, all bodies will move with the greatest freedom; and the planets and comets will constantly pursue their revolutions in orbits given in kind and position, according to the laws above explained; but though these bodies may, indeed, continue in their orbits by the mere laws of gravity, yet they could by no means have at first derived the regular position of the orbits themselves from those laws.
>
> The six primary planets are revolved about the sun in circles concentric with the sun, and with motions directed towards the same parts, and almost in the same plane. Ten moons are revolved about the earth, Jupiter, and Saturn, in circles concentric with them, with the same direction of motion, and nearly in the planes of the orbits of those planets; but it is not to be conceived that mere mechanical causes could give birth to so many regular motions, since the comets range over all parts of the heavens in very eccentric orbits; for by that kind of motion they pass easily through the orbs of the planets, and with great rapidity; and in their aphelions, where they move the slowest, and are detained the longest, they recede to the greatest

distances from each other, and hence suffer the least disturbance from their mutual attractions. This most beautiful system of the sun, planets, and comets, could only proceed from the counsel and dominion of an intelligent and powerful Being. And if the fixed stars are the centres of other like systems, these, being formed by the like wise counsel, must be all subject to the dominion of One; especially since the light of the fixed stars is of the same nature with the light of the sun, and from every system light passes into all the other systems; and lest the systems of the fixed stars should, by their gravity, fall on each other, he hath placed those systems at immense distances from one another.

Newton's God is not merely a "philosophical" God, the impersonal and uninterested First Cause of the Aristotelians, or the — for Newton — utterly indifferent and world-absent God of Descartes. He is — or, in any case, Newton wants him to be — the Biblical God, the effective Master and Ruler of the world created by him: [6]

This Being governs all things, not as the soul of the world, but as Lord over all; and on account of his dominion he is wont to be called *Lord God* παντοκράτωρ or *Universal Ruler*; for *God* is a relative word, and has a respect to servants; and *Deity* is the dominion of God not over his own body, as those imagine who fancy God to be the soul of the world, but over servants. The Supreme God is a Being eternal, infinite, absolutely perfect; but a being, however perfect, without dominion, cannot be said to be Lord God; for we say, my God,, your God, the God of *Israel*, the God of Gods, the Lord of Lords; but we do not say, my Eternal, your Eternal, the Eternal of *Israel*, the Eternal of Gods; we do not say, my Infinite, or my Perfect: these are titles which have no respect to servants. The word God usually signifies *Lord*; but every Lord is not a God. It is

> the dominion of a spiritual being which constitutes a God: a true, supreme, or imaginary God. And from this true dominion it follows that the true God is a living, intelligent, and powerful Being; and from his other perfections, that he is supreme, or most perfect. He is eternal and infinite, omnipotent and omniscient; that is, his duration reaches from eternity to eternity; his presence from infinity to infinity; he governs all things, and knows all things that are or can be done.

His duration reaches from eternity to eternity; his presence from infinity to infinity . . . the Newtonian God is, patently, not above time and space: His eternity is sempiternal duration, His omnipresence is infinite extension. This being so, it is clear why Newton insists: [7]

> He is not eternity and infinity, but eternal and infinite; he is not duration or space, but he endures and is present.

And yet, like the God of Henry More and of Joseph Raphson, he not only "endures forever and is everywhere present"; but it is "by existing always and everywhere" that "he constitutes duration and space." It is not surprising therefore that [8]

> since every particle of space is *always*, and every indivisible moment of duration is *everywhere*, certainly the Maker and Lord of all things cannot be *never* and *nowhere*. Every soul that has perception is, though in different times and in different organs of sense and motion, still the same indivisible person. There are given successive parts in duration, coexistent parts in space, but neither the one nor the other in the person of a man, or his thinking principle; and much less can they be found in the thinking substance of God. Every man, so far as he is a thing that has perception, is

one and the same man during his whole life, in all and each of his organs of sense.

And that,[9]

He is omnipresent not *virtually* only, but also *substantially*; for virtue cannot subsist without substance. In him are all things contained and moved; yet neither affects the other: God suffers nothing from the motion of bodies; bodies find no resistance from the omnipresence of God. It is allowed by all that the Supreme God exists necessarily; and by the same necessity he exists *always* and *everywhere*.

Thus "in Him we live, we move and we are," not metaphorically or metaphysically as St. Paul meant it, but in the most proper and literal meaning of these words. We — that is, the world — are in God; in God's space, and in God's time. And it is because of this ubiquitous and sempiternal co-presence with things that God is able to exercise His dominion upon them; and it is this dominion or, more exactly, the effect of this dominion that reveals to us His otherwise unknowable and incomprehensible essence: [10]

We know him only by his most wise and excellent contrivances of things, and final causes; we admire him for his perfections; but we reverence and adore him on account of his dominion: for we adore him as his servants; and a god without dominion, providence, and final causes, is nothing else but Fate and Nature. Blind metaphysical necessity, which is certainly the same always and everywhere, could produce no variety of things. All that diversity of natural things which we find suited to different times and places could arise from nothing but the ideas and will of a Being necessarily existing. But, by way of allegory, God is said to see, to speak, to laugh, to love, to hate, to desire, to give, to receive, to rejoice, to be angry, to fight, to frame, to work, to build; for all our notions of God are taken from the ways of mankind by a certain similitude, which, though not perfect, has some likeness, however. And thus much concerning God; to discourse of whom from the appearances of things, does certainly belong to Natural Philosophy.

Thus much for God; or for Berkeley. As for gravity, or for Leibniz, Newton explains that he does not introduce into philosophy "occult qualities" and magical causes, but, on the contrary, restricts his investigation to the study and analysis of observable, patent phenomena, renouncing, at least for the time being, the causal explanation of the experientially and experimentally established laws: [11]

Hitherto we have explained the phenomena of the heavens and of our sea by the power of gravity, but have not yet assigned the cause of this power. This is certain, that it must proceed from a cause that penetrates to the very centres of the sun and planets, without suffering the least diminution of its force; that operates not according to the quantity of the surfaces of the particles upon which it acts (as mechanical causes used to do), but according to the quantity of the solid matter which they contain, and propagates its virtue on all sides to immense distances, decreasing always as the inverse square of the distances. . . . But hitherto I have not been able to discover the cause of those properties of gravity from phenomena, and I feign no hypotheses; for whatever is not deduced from the phenomena is to be called an hypothesis; and hypotheses, whether metaphysical or physical, whether of occult qualities or mechanical, have no place in experimental philosophy. In this philosophy particular propositions are inferred from the phenomena, and afterwards rendered general by induction. Thus it was that the impenetrability, the mobility, and the impulsive force of bodies, and the laws of motion and of gravitation, were discovered. And to us it is enough that gravity does really exist, and act according to the laws which we have explained, and abundantly serves to account for all the motions of the celestial bodies, and of our sea.

"I feign no hypotheses . . ."[12] *Hypotheses non fingo* . . . a phrase that became extremely famous and also like all, or nearly all, celebrated utterances torn out of their context, completely perverted in its meaning. "I feign no hypotheses." Of course not; why should Newton "feign hypotheses," that is, fictitious and fanciful conceptions not deduced from phenomena and having therefore no basis in reality? Hypotheses, "whether of occult qualities or mechanical have no place in experimental philosophy"—of course not, as this kind of hypothesis is, by definition, either false or at least unable to conduce to experiments and be checked and confirmed (or disproved) by them. Gravity is not a hypothesis, or an "occult" quality. The existence of gravity, insofar as it is a statement about the behaviour of bodies, or about the existence of centripetal forces in consequence of which bodies, instead of moving in straight lines (as they should, according to the principle or law of inertia), are deflected and move in curves, is a patent fact; the identification of the cosmical "force" which determines the motion of planets with that in consequence to which bodies fall, that is, move towards the center of the earth, is certainly an important discovery. But the assumption of the existence *in* bodies of a certain force which enables them to act upon other bodies and to *attract* them is not a hypothesis either. Not even one that makes use of occult qualities. It is mere and pure nonsense.

As for " mechanical " hypotheses, that is, *those of Descartes, Huygens and Leibniz,* they have no place in experimental philosophy simply because they attempt to do something that cannot be done, as Newton hints rather broadly, indeed at the very beginning of the *General Scholium* where he shows that " the hypothesis of vortices is pressed with many difficulties." Mechanical — feigned — hypotheses, as his pupil and editor Roger Cotes explains in his famous preface to the second edition of the *Principia,* are the special and favorite dish of the Cartesians, who, moreover, are conduced by them into the assumptions of truly and really " occult " properties and realities. Thus having explained the sterility of Aristotelian and scholastic philosophy of nature, Cotes continues: [13]

> Others have endeavored to apply their labors to greater advantage by rejecting that useless medley of words [of the scholastic natural philosophy]. They assume that all matter is homogeneous, and that the variety of forms which is seen in bodies arises from some very plain and simple relations of the component particles. And by going from simple things to those which are more compounded they certainly proceed right, if they attribute to those primary relations no other relations than those which Nature has given. But when they take a liberty of imagining at pleasure unknown figures and magnitudes, and uncertain situations and motions of the parts, and moreover of supposing occult fluids, freely pervading the pores of bodies, endued with an all-performing subtility, and agitated with occult motions, they run out into dreams and chimeras, and neglect the true constitution of things, which certainly is not to be

derived from fallacious conjectures, when we can scarce reach it by the most certain observations. Those who assume hypotheses as first principles of their speculations, although they afterwards proceed with the greatest accuracy from those principles, may indeed form an ingenious romance, but a romance it will still be.

As for Leibniz, whom Cotes does not mention by name, yet clearly, though somewhat parodistically, hints at, he is no better than the Cartesians; or perhaps even worse, as he assumes the existence around " the comets and planets . . . of atmospheres . . . which by their own nature move around the sun and describe conic sections " (an unmistakable allusion to the " harmonic circulation " of the great German mathematician and arch-foe of Newton), a theory which Cotes declares to be a " fable " as fantastic as that of the Cartesian vortices, and of which he presents a rather witty and biting parody: [14]

Galileo has shown that when a stone projected moves in a parabola, its deflection into that curve from its rectilinear path is occasioned by the gravity of the stone towards the earth, that is, by an occult quality. But now somebody, more cunning than he, may come to explain the cause after this manner. He will suppose a certain subtile matter, not discernible by our sight, our touch, or any other of our senses, which fills the spaces which are near and contiguous to the surface of the earth, and that this matter is carried with different directions, and various, and often contrary, motions, describing parabolic curves. Then see how easily he may account for the deflection of the stone above spoken of. The stone, says he, floats in this subtile fluid, and following its motion, can't choose but describe the same figure. But the fluid moves in parabolic curves, and therefore the stone must move in a parabola, of course. Would not the acuteness of this philosopher be thought very extraordinary, who could deduce the appearances of Nature from mechanical causes, matter and motion, so clearly that the meanest man may understand it? Or indeed should not we smile to see this new *Galileo* taking so much mathematical pains to introduce occult qualities into philosophy, from whence they have been so happily excluded? But I am ashamed to dwell so long upon trifles.

Trifles? As a matter of fact, we are not dealing with trifles. The use of " hypotheses " constitutes, indeed, a deep and dangerous perversion of the very meaning and aim of natural philosophy: [15]

The business of true philosophy is to derive the natures of things from causes truly existent, and to inquire after those laws on which the Great Creator actually chose to found this most beautiful Frame of the World, not those by which he might have done the same, had he so pleased.

Yet the partisans of mechanical hypotheses, that is, once more, the Cartesians — and Leibniz — not only forget this fundamental rule, they go much farther and, by the denial of void space as impossible, they impose upon God a certain determinate manner of action, restrict his power and freedom, and subject him, thus, to necessity; finally, they deny altogether that the world was freely created by God. A teaching not only infamous, but also false (as Newton has shown): [16]

Therefore they will at last sink into the mire of that infamous herd who dream that all things are governed by fate and not by providence, and that matter exists by the necessity of its nature always and everywhere, being infinite and eternal. But supposing these things, it must be also everywhere uniform; for variety of forms is entirely inconsistent with necessity. It must be also unmoved; for if it be necessarily moved in any determinate direction, with any determinate velocity, it will by a like necessity be moved in a different direction with a different velocity; but it can never move in different directions with different velocities; therefore it must be unmoved. Without all doubt this world, so diversified with that variety of forms and motions we find in it, could arise from nothing but the perfectly free will of God directing and presiding over all.

From this fountain it is that those laws, which we call the laws of Nature, have flowed, in which there appear many traces indeed of the most wise contrivance, but not the least shadow of necessity. These therefore we must not seek from uncertain conjectures, but learn them from observations and experiments. He who is presumptuous enough to think that he can find the true principles of physics and the laws of natural things by the force alone of his own mind, and the internal light of his reason, must either suppose that the world exists by necessity, and by the same necessity follows the laws proposed; or if the order of Nature was established by the will of God, that himself, a miserable reptile, can tell what was fittest to be done. All sound and true philosophy is founded on the appearance of things; and if these phenomena inevitably draw us, against our wills, to such principles as most clearly manifest to us the most excellent counsel and supreme dominion of the All-wise and Almighty Being, they are not therefore to be laid aside because some men may perhaps dislike them. These men may call them miracles or occult qualities, but names maliciously given ought not to be a disadvantage to the things themselves, unless these men will say at last that all philosophy ought to be founded in atheism. Philosophy must not be corrupted in compliance with these men, for the order of things will not be changed.

We see now clearly why we must not feign hypotheses. Hypotheses, especially mechanical ones, implying the rejection of void space and the assertion of infinity and therefore of the necessity of matter, are not only false; they lead straight away towards atheism.

Mechanical hypotheses concerning gravity, as a matter of fact, deny God's action in the world and push him out of it. It is indeed, practically certain — and this knowledge makes the " feigning of hypotheses " completely nonsensical — that the true and ultimate cause of gravity is the action of the " spirit " of God. Newton therefore concludes his *General Scholium:* [17]

> And now we might add something concerning a certain most subtle spirit which pervades and lies hid in all gross bodies; by the force and action of which spirit the particles of bodies attract one another at near distances, and cohere, if contiguous; and electric bodies operate to greater distances, as well repelling as attracting the neighbouring corpuscles; and light is emitted, reflected, refracted, inflected, and heats bodies; and all sensation is excited, and the members of animal bodies move at the command of the will, namely, by the vibrations of this spirit, mutually propagated along the solid filaments of the nerves, from the outward organs of sense to the brain, and from the brain into the muscles. But these are things that cannot be explained in few words, nor are we furnished with that sufficiency of experiments which is required to an accurate determination and demonstration of the laws by which this electric and elastic spirit operates.

XI. The Work-Day God and the God of the Sabbath

Newton & Leibniz

Newton's veiled and Roger Cotes' open counterattack upon the "plenists" did not remain unanswered. If the Cartesians, properly speaking, did not react, Leibniz, in a letter to the Princess of Wales,[1] written in November 1715, replied to the accusations formulated by Cotes by expressing to his august correspondent his misgivings concerning the weakening of religion and the spread of materialism and godless philosophies in England, where some people attributed materiality not only to souls but even to God, where Mr. Locke doubted the immateriality and the immortality of the soul, and where Sir Isaac Newton and his followers professed rather low and unworthy ideas about the power and wisdom of God. Leibniz wrote: [2]

> Sir *Isaac Newton* says, that Space is an *Organ*, which God makes use of to perceive Things by. But if God stands in need of any *Organ* to perceive Things by, it will follow, that they do not depend altogether upon him, nor were produced by him.

> Sir *Isaac Newton*, and his Followers, have also a very odd Opinion concerning the Work of God. According to their Doctrine, God Almighty wants to *wind up* his Watch from Time to Time: Otherwise it would cease to move. He had not, it seems, sufficient Foresight to make it a perpetual Motion. Nay, the Machine of God's making, is so imperfect, according to these Gentlemen, that he is obliged to *clean* it now and then by an extraordinary Concourse, and even to *mend* it, as a Clockmaker mends his Work; who must consequently be so much the more unskilful a Workman, as he is often obliged to mend his Work and to set it Right. According to *My* Opinion, the *same* Force and Vigour remains always in the World, and only passes from one part of Matter to another, agreeably to the Laws of Nature, and the beautiful pre-established Order.

An accusation of the kind formulated by Leibniz could not, of course, be left without refutation. Yet, as it was obviously below the dignity and standing of Sir Isaac — who, moreover, hated all polemics and public discussions — to do it himself, the task fell upon the shoulders of Dr. Samuel Clarke, the faithful pupil and friend of Newton, who translated his *Opticks* into Latin,[3] and, as far back as 1697, stuffed with Newtonian footnotes his translation of Rohault's Cartesian *Physics.* A long-drawn-out and extremely interesting correspondence resulted, which ended only with the death of Leibniz, and which throws a vivid light upon the conflicting positions of the two philosophers (Leibniz and Newton) as well as upon the fundamental issues that were in question.

Thus, Dr. Clarke, though recognizing the deplorable fact that there were, in England as elsewhere, persons who denied even natural religion or corrupted it entirely, explained that it was due to the spread of false materialistic philosophies (which were also responsible for the materialization of the soul and even God, mentioned by Leibniz); pointed out that these people were most effectively combatted by the mathematical philosophy, the only philosophy which proves that matter is the smallest and the least important part of the universe.[4] As for Sir Isaac Newton, he does not say that space is an organ which God uses in order to perceive things, nor that God needs any means for perceiving them. Quite the contrary, he says that God, being everywhere, perceives them by his immediate presence in the very space where they are. And it is just in order to explain the immediacy of this perception that Sir Isaac Newton — comparing God's perception of *things* with the mind's perception of *ideas* — said that infinite space is, so to speak, as the *sensorium* of the Omnipresent God.[4a]

From the point of view of the Newtonian, Leibniz's reproach of belittling God's power and wisdom by obliging Him to repair and to wind up the world clock is both unfair and unjustified; on the contrary, it is just by His constant and vigilant action, by conferring on the world new energy that prevents its decay into chaotic disorder and immobility, that God manifests His presence in the world and the blessing of His providence. A Cartesian, or a Leibnizian God, interested only in conserving in its being a mechanical clockwork set once and forever, and endowed, once and forever with a constant amount of energy, would be nothing better than an absent God. Clarke therefore states rather wickedly that the assimilation of the world to a perfect mechanism moving without God's intervention,[5]

. . . is the Notion of *Materialism* and *Fate*, and tends (under pretence of making God a *Supra-Mundane Intelligence*) to exclude *Providence* and *God's Government* in reality out of the World. And by the same Reason that a *Philosopher* can represent all Things going on from the beginning of the Creation, *without* any Government or Interposition of Providence, a *Sceptick* will easily Argue still farther Backwards, and suppose that Things have from Eternity gone on (as they now do) *without* any true Creation or Original Author at all, but only what such Arguers call *All-Wise and Eternal Nature.* If a *King* had a *Kingdom* wherein all Things would continually go on *without* his Government or Interposition, or *without* his Attending to and Ordering what is done therein; It would be to *him*, merely a *Nominal* Kingdom; nor would he in reality deserve at all the Title of King or Governor. And as those Men, who pretend that in an Earthly Government Things may go on perfectly well *without* the *King himself* ordering or disposing of any Thing, may reasonably be suspected that they would like very well to set the King aside: so whosoever contends, that the Course of the World can go on *without* the Continual direction of *God,* the Supreme Governor; his Doctrine does in Effect tend to exclude God out of the World.

Confronted with Dr. Clarke's reply that rather unexpectedly placed him under the obligation to defend himself against Clarke's sly insinuations, Leibniz struck back by pointing out that " mathematical " principles are not opposed to, but identical with, those of materialism and have been claimed by Democritus and Epicurus as well as by Hobbes; that the problem dealt with is not a mathematical but a metaphysical one, and that metaphysics, in contradistinction to mere mathematics, has to be based on the *principle of sufficient reason*; that this principle, applied to God, necessarily implies the consideration of God's wisdom in planning and creating the universe, and that, *vice versa*, the neglect of this principle (Leibniz does not *say* so outright, yet he suggests that such is the case of the Newtonians) leads directly to the world-view of Spinoza, or, on the other hand, to a conception of God closely resembling that of the Socinians,[5a] whose God is so utterly lacking in foresight that He has "to live from day to day." The Newtonians point out that, according to them, and in contradistinction to the materialists, matter is the least important part of the universe, which is chiefly constituted by void space. But after all, Democritus and Epicurus admitted void space just as Newton does, and if they differed from him in believing that there was much more matter in the world than there is according to Newton, they were in this respect preferable to the latter; indeed, more matter means more opportunities for God to exercise His wisdom and power, and that is a reason, or at least one of the reasons, why, in truth, there is no void space at all in the universe, and that space is everywhere full of matter.

But to come back to Newton. In spite of all the explanations of his friends,[6]

> I find [writes Leibniz] in express Words, in the *Appendix* to Sir *Isaac Newton's Opticks*, that *Space* is the *Sensorium* of *God*. But the Word *Sensorium* hath always signified the *Organ* of Sensation. He, and his Friends, may *now*, if they think fit, explain themselves quite otherwise: I shall not be against it.

And as for the accusation of making the world a self-sufficing mechanism and reducing God to the status of an *intelligentia supra-mundana,* Leibniz replies that he never did so, that is, that he never denied that the created world needed God's continuous concourse, but only asserted that the world is a clock that does not need mending, since, before creating it, God saw, or foresaw, everything; and that he never excluded God from the world, though he did not, as his adversaries seem to do, transform Him into the soul of the world. Indeed, if God has, from time to time, to correct the natural development of the world, he can do it either by supernatural means, that is, by a miracle (but to explain natural things and processes by miracles is absurd); or He can do it in a *natural* way: in this case God is included in nature and becomes *anima mundi.* Finally,[7]

> The comparison of a King, under whose Reign every thing should go on without his Interposition, is by no means to the present Purpose; since God preserves every thing continually, and nothing can subsist without him. His Kingdom therefore is not a *Nominal* one.

Otherwise we should have to say that a Prince who has so well educated his subjects that they never infringe his laws is a Prince only in name.

Leibniz does not express, as yet, his ultimate objections to Newton; the fundamental opposition appears nevertheless pretty clearly: the God of Leibniz is not the Newtonian Overlord who makes the world as he wants it and continues to act upon it as the Biblical God did in the first six days of Creation. He is, if I may continue the simile, the Biblical God on the Sabbath Day, the God who has finished his work and who finds it good, nay, the very best of all possible worlds, and who, therefore, has no more to act upon it, or in it, but only to conserve it and to preserve it in being. This God is, at the same time — once more in contradistinction to the Newtonian one — the supremely rational Being, the principle of sufficient reason personified, and for this very reason, He can act only according to this principle, that is, only in order to produce the greatest perfection and plenitude. He cannot therefore — any more than the God of Giordano Bruno with whom (in spite of His being a mathematician and a scientist) He has a great deal in common — either make a finite universe, or suffer void space either inside or outside the world.

It is hardly surprising that, having read Leibniz's answer to his criticism, Dr. Clarke felt himself compelled to reply: Leibniz's hints were too damaging,[8] his tone too superior, and, moreover, his insistence on the implications of the term "*sensorium*," somewhat hastily and perhaps unhappily used by Newton, far too menacing to allow Clarke to leave Leibniz in the position of having had the last word.

Starting thus from the beginning, Clarke explains [9] that the "principles of mathematical philosophy" are by no means identical with, but radically opposed to, those of materialism, precisely in that they deny the possibility of a purely naturalistic explanation of the world and postulate — or demonstrate — its production by the purposeful action of a free and intelligent Being. And as for Leibniz's appeal to the principle of sufficient reason, it is true that nothing exists without sufficient reason: where there is no cause, there is also no effect; yet the said sufficient reason can be simply the will of God. Thus, for instance, if one considers why a system, or a certain piece, of matter is created in one place, and another one in another, and not *vice versa,* there can be no other reason for that than the pure will of God. It it were not so — that is, if the principle of sufficient reason were taken absolutely, as Leibniz does — and if this will could never act unless predetermined by some cause, as a balance cannot move unless some weight make it turn, God would have no liberty of choice, which would be replaced by necessity.

As a matter of fact, Dr. Clarke subtly suggests that Leibniz, indeed, deprives his God of all liberty. Thus he forbids him to create a limited quantity of matter . . . yet by the same argument one could prove that the number of men or of any kind of creatures whatsoever should be infinite (which, of course, would imply the eternity and necessity of the world).

As for the (Newtonian) God, he is neither an *intelligentia mundana,* nor an *intelligentia supra-mundana;* nor is he an *anima mundi,* but an intelligence which is everywhere, in the world and outside it, in everything, and above everything. And he has no organs as Leibniz persists in insisting.[10]

> The Word *Sensory* does not properly signify the *Organ,* but the *Place* of Sensation. The *Eye,* the *Ear,* &c. are Organs, but not *Sensoria.*

Moreover, Newton does not say that place *is* a *sensorium,* but calls it thus only by way of comparison, in order to indicate that God really and effectively perceives things in themselves, where they are, being present to them, and not purely transcendent — present, acting, forming and reforming (which last term, just as the term " correcting," must be understood in respect to us, or to God's works, not indeed as implying change in God's designs): thus if [11]

> the present Frame of the Solar System (for instance) according to the present Laws of Motion, will in time *fall into Confusion*; and perhaps, after That, will be *amended* or put into a *new Form*

it will be new in respect to us, or to itself, not new in respect to God whose eternal plan implied just such an intervention in the normal course of events. To forbid God to do that, or to declare all God's action in the world to be miraculous or supernatural, means excluding God from the government of the world. It may be, concedes Clarke, that in this case He would still remain its Creator; He would certainly no longer be its governor.

The second paper of Dr. Clarke made Leibniz angry. Why, he complains, did they grant me this important principle that *nothing happens without a sufficient reason why it should be so rather than otherwise*, but they grant it only in words, not in fact. Moreover, they use against me one of my own demonstrations against *real absolute space*, that idol (in the sense of Bacon) of some modern Englishmen. Leibniz is right, of course: to say, as Clarke does, that God's will is, as such, a sufficient reason for anything, is to reject the principle, and to reject also the thorough-going rationalism which supports it. And to use the conception of homogeneous, infinite, real space as a basis for the demonstration that God's free (that is, unmotivated, irrational) will can, and must, be considered as a " sufficient reason " for something, is to insult the intelligence; and to force Leibniz to discuss the problem of space (something he did not very much want to do): [12]

> These Gentlemen maintain therefore, that *Space* is a *real absolute Being.* But this involves them in great difficulties; for such a *Being* must needs be *Eternal* and *Infinite.* Hence Some have believed it to be *God himself,* or, one of his Attributes, his Immensity. But since Space consists of *Parts,* it is not a thing which can belong to God.

All that, as we know, is perfectly true. Nevertheless Leibniz's criticism of the Newtonian or, more generally, the absolutist conception of space, forgets that those who hold it deny that space consists of parts — *partes extra partes* — and assert, on the contrary, that it is indivisible. Leibniz is perfectly right, too, in asserting that [13]

> *Space* is Something absolutely *Uniform;* and, without the Things placed in it, *One Point* of Space does not absolutely differ in any respect whatsoever from *Another Point* of Space. Now from hence it follows, (supposing Space to be Something in it self, besides *the Order of Bodies among themselves,*) that it is impossible there should be a Reason, why God, preserving the same Situations of Bodies among themselves, should have placed them in Space after *one certain particular manner,* and not otherwise; why every thing was not placed the *quite contrary way,* for instance, by changing East into West.

Yet the conclusions drawn by Leibniz and by Clarke from the same, hypothetically admitted facts are diametrically opposed. Leibniz believes that in this case, that is, in the absence of reasons for choice, God would not be able to act; and *vice versa*, from the fact of the choice and of acting, he deduces the rejection of the fundamental hypothesis, that is, the existence of an absolute space, and proclaims that space, like motion, is purely relative, or even more, is nothing else but the order of coexistence of bodies and would not exist if there were none, just as time is nothing else but the order of succession of things and events, and would not exist if there were no things or events to be ordered.

The Newtonian, on the other hand, concludes the freedom of God, that is, the non-necessity of a determining reason or motive for God's choice and action. For Leibniz, of course, this unmotivated choice is vague indifference, which is the contrary of true freedom; but for the Newtonian, it is the absolutely motivated action of the Leibnizian God which is synonymous with necessity.

The Newtonians assert that, left to itself, the motive force of the universe would decrease and finally disappear. But, objects Leibniz,[14]

> if *active Force* should *diminish* in the Universe, by the Natural Laws which God has established; so that there should be need for him to give a *new Impression* in order to restore that Force, like an Artist, Mending the Imperfections of his Machine; the Disorder would not only be with respect to *Us*, but also with respect to *God himself*. He *might have* prevented it and taken better Measures to avoid such an Inconvenience: And therefore, indeed, he has actually done it.

The Newtonians protest against Leibniz's assertion that they make nature a perpetual miracle. And yet, if God wanted to make a free body revolve around a fixed center, though not acted upon by any other creature, He would not be able to achieve it without a miracle since such a motion cannot be explained by the nature of bodies. For a free body naturally moves away from a curved line along its tangent. Thus mutual attraction of bodies is something miraculous as it cannot be explained by their nature.

From now on the discussion broadens and deepens. The " papers " become longer and longer. The skirmish develops into a pitched battle. Leibniz and Clarke go at each other hammer and tongs. It is true that, to a large extent, they simply repeat, or elaborate, the same arguments — philosophers, I have already said it, seldom, if ever, convince each other, and a discussion between two philosophers resembles as often as not a " dialogue de sourds " — and yet they come more and more into the open, and the fundamental issues come more and more to the foreground.

Thus, for instance, in his *third paper*, Dr. Clarke re-objects to Leibniz that it is preposterous to subject God to the law of strict motivation and to deprive Him of the faculty of making a choice between two identical cases. Indeed, when God creates a particle of matter in one place rather than in another, or when He places three identical particles in a certain order rather than in another, He cannot have any reason for doing so except His pure will. The perfect equivalence of the cases, a consequence of the identity of material particles and of the isomorphism of space, is no more a reason for denying God's freedom of choice than it is an objection to the existence of an absolute, real and infinite space. And as for its relation to God, misrepresented by Leibniz, Clarke states the correct, Newtonian, that is, More's, doctrine: [15]

> *Space* is not a *Being*, an eternal and infinite *Being*, but a *Property* [attribute], or a consequence of the Existence of a Being infinite and eternal. *Infinite Space*, is *Immensity*. But *Immensity* is *not God*: And therefore Infinite Space, is not God. Nor is there any Difficulty in what is here alleged about Space having *Parts*. For Infinite Space is One, absolutely and essentially indivisible: And to suppose it *parted*, is a contradiction in Terms; because there must be Space in the *Partition it self*; which is to suppose it *parted*, and yet *not parted* at the same time. The *Immensity* or *Omnipresence* of God, is no more a dividing of his Substance into *Parts*; than his *Duration*, or continuance of existing is a dividing of his existence into *Parts*. There is no difficulty here, but what arises from the *figurative* Abuse of the Word, *Parts*.

It is not Newton's admission, it is Leibniz's denial, of absolute space that leads to difficulties and absurdities. Indeed, if space were only relative, and nothing but the order and arrangement of things, then a mere displacement of a system of bodies from one place to another (for instance, of our world to the region of the farthest fixed stars) would be no change at all, and it would follow therefrom that the two places would be the same place. . . .[16] It would follow also that, if God should move the whole world in a straight line, then, whatever the speed of this motion, the world would remain in the same place, and that nothing would happen if that motion were suddenly stopped.[17]

And if *time* were only an order of succession, then it would follow that, if God had created the world some millions of years earlier, it would, nevertheless, have been created at the same time.

We shall see in a moment what Leibniz has to object to in Dr. Clarke's reasonings (he will find them meaningless); as for us, we have to admit that they are by no means as absurd as may seem at first glance; they only represent, or imply, a formal breach (already accomplished by Henry More) with the main philosophico-theological tradition to which Leibniz remains fundamentally faithful: the Newtonians, as we know, do not attach time and space to creation but to God, and do not oppose God's eternity and immensity to sempiternity and spatial infinity, but, on the contrary, identify them. Clarke thus explains: [18]

> *God*, being *Omnipresent*, is really *present* to everything, *Essentially* and *Substantially*. His Presence *manifests* it self indeed by its *Operation*, but it could not operate if it was not *There*.

Nothing, indeed, can act without being *there*; not even God: there is no action at a distance; not even for God. Yet as God is everywhere " there," He can, and does, act everywhere, and therefore, Leibniz's assertion to the contrary notwithstanding, He can achieve without miracle, but by His own — or some creature's — action that a body be deflected from the tangent and can even make a body turn around a fixed center instead of running away along the tangent; whether God in order to produce this effect acts Himself, or through a creature, is of no avail: in neither case would it be a miracle as Leibniz pretends.

It is clear that, for Clarke, Leibniz's assertion — as well as his rejection as " imperfection " of the diminution of the moving power in the world — is based on the assumption of the necessary self-sufficiency of nature; a conception, as we know, utterly unacceptable for the Newtonians who see in it a means of excluding God from the world.

But let us come back to Clarke's objection to Leibniz's conception of space. The first argument of Samuel Clarke is not very good, as the displacement imagined by him would be not only absolute but also relative to the aggregate of the fixed stars. But the second one is perfectly valid: in the infinite universe of Newtonian physics any, and every, body can be considered as possessing — or not possessing — a uniform, rectilinear motion in a certain direction, and though the two cases would be perfectly indistinguishable one from another, the passage from the one to the other would be accompanied by very determined effects. And if the motion were not uniform but accelerated, we should even be able to perceive it (something that would not happen if motion and space were only relative): all that is an inevitable consequence of the Newtonian principle of inertia.

Clarke, of course, does not stop here. For him — as for Bentley or Raphson — the radical distinction of matter and space implies the belief in the possible and perhaps even real finitude of the universe. Why, indeed, should matter, which occupies so small a part of space, be infinite? Why should we not admit, on the contrary, that God has created a determined amount of it, just as much as was needed for this very world, that is, for the realization of the aims that God had in creating it?

The *fourth* paper of Leibniz leads us directly to the deepest metaphysical problems. Leibniz starts by asserting with the utmost energy the absolute panarchy of the principle of sufficient reason: no action without choice, no choice without determining motive, no motive without a difference between the conflicting possibilities; and therefore — an affirmation of overwhelming importance — no two identical objects or equivalent situations are real, or even possible, in the world.[19]

As for space, Leibniz reasserts just as vigorously that space is a function of bodies and that, where there are no bodies, there is also no space.[20]

> The same reason, which shows that *extra-mundane* Space is *imaginary*, proves that *All empty Space* is an *imaginary* thing; for they differ only as greater and less.

This does not mean, of course, that, according to Leibniz, the world and space are both limited in extension, as was thought by the mediaeval philosophers who spoke about the " imaginary " space " outside " of the world; but, on the contrary, that void space, be it outside or inside the world, is pure fiction. Space, everywhere, is full; indeed,[21]

> There is no *possible* Reason, that *can limit* the quantity of Matter; and therefore such limitation can have no place.
>
> Now, let us fancy a *Space* wholly *empty*, God *could* have placed some Matter in it, without derogating in any respect from all other things; Therefore he hath actually placed some Matter in That Space: Therefore, there is no Space wholly *Empty*: Therefore All is full.[22] The same Argument proves that there is no Corpuscle, but what is Subdivided.[23]

Moreover, the idea of void space is a metaphysically impossible idea, against which Leibniz erects objections analogous to, and probably derived from, those that Descartes opposed to Henry More: [24]

> If Space is a property or Attribute, it must be the Property of some *Substance*. But *what Substance* will that *Bounded* empty Space be an Affection or Property of, which the Persons I am arguing with, suppose to be between Two Bodies?

This is a reasonable question, but a question to which Henry More had already given an answer, which Leibniz however chooses to disregard; he continues therefore: [25]

> If *Infinite Space* is *Immensity, finite Space* will be the Opposite to Immensity, that is, 'twill be *Mensurability*, or *limited Extension*. Now Extension must be the Affection of some thing extended. But if That Space be empty, it will be an Attribute *without a Subject*, an Extension without any thing extended. Wherefore by making Space a *Property*, the Author falls in with My Opinion, which makes it an Order of things, and not any thing absolute.

By no means; of course there is no attribute without substance; but as we know, for the "Author" that substance is God. Leibniz does not admit it, and develops the awkward consequences of the absolutist conception: [26]

> If Space is an absolute *reality*; far from being a *Property* or an Accident opposed to Substance, it will have a *greater reality* than *Substances* themselves. God cannot destroy it, nor even change it in any respect. It will be not only immense in the whole, but also *Immutable* and *Eternal* in every part. There will be an infinite number of Eternal things besides God.

As we know, it is just what the Newtonians, or the Henry

More-ists assert, denying, of course, that space is something "besides" God. But their teaching, according to Leibniz, implies contradictions: [27]

> To say that *Infinite Space* has no *Parts*, is to say that it does not consist of *finite* Spaces; and that Infinite Space might subsist, though all finite Spaces should be reduced to nothing. It would be as if one should say, in the *Cartesian* Supposition of a material extended unlimited World that such a World might subsist, though all the Bodies of which it consists, should be reduced to nothing.

By no means; Leibniz does not understand the difference between his own conception of space — a lattice of quantitative relations — and that of Newton, for whom space is a unity which precedes and makes possible all relations that can be discovered in it. Or, more probably, since it is rather difficult to believe that there was something that Leibniz did not understand, he *does* understand, but does not admit the conception of Newton. Thus he writes: [28]

> If *Space* and *Time* were anything absolute, that is, if they were any thing else, besides certain *Orders* of Things; then indeed my assertion would be a *Contradiction*. But since it is not so, the Hypothesis [*that Space and Time are any thing absolute*] is contradictory, that is 'tis an impossible Fiction.

As for the examples and counter-objection of Dr. Clarke, Leibniz deals with them in a rather off-hand manner. Thus he reasserts that those who fancy that the active powers decrease by themselves in the world do not know the principal laws of nature; that to imagine God moving the world in a straight line is to compel him to do something wholly meaningless, an action without rime or reason, that is, an action that it is impossible to attribute to God. Finally, concerning attraction, which Clarke endeavors to present as something natural, Leibniz repeats: [29]

> 'Tis also a supernatural thing, that Bodies should *attract* one another at a distance, without any intermediate Means; and that a Body should move round, without receding in the Tangent, though nothing hinders it from so receding. For these Effects cannot be explained by the Nature of things.

Leibniz's repeated appeal to the principle of sufficient reason did not, needless to say, convince or even appease Clarke. Quite the contrary: it seemed to him to confirm his worst apprehensions. In the *fourth* reply he writes: [30]

> This Notion leads to universal *Necessity and Fate*, by supposing that *Motives* have the same relation to the *Will of an Intelligent Agent*, as *Weights* have to a *Balance*; so that of *two* things absolutely indifferent, an Intelligent Agent can no more choose *Either*, than a Balance can move it self when the Weights on both sides are Equal. But the Difference lies here

in the distinction, disregarded by Leibniz, between a free and intelligent being, who is a self-determining agent, and a mere mechanism, which, in the last analysis, is always passive. If Leibniz were right about the impossibility of a plurality of identical objects, no creation would ever have been possible; matter, indeed, has one identical nature, and we can always suppose that its parts have the same dimension and figure.[31] In other terms: the atomic theory is utterly incompatible with Leibniz's conception; which is, of course, perfectly true. For Leibniz there cannot be in the world two identical objects; moreover Leibniz, like Descartes, denies the existence of last, indivisible, hard particles of matter, without which Newtonian physics is inconceivable.

Leibniz's linking space (and time) with the world, and his assertion of the fictitious (imaginary) character of void space and "void" time seem to Clarke utterly unreasonable; and also full of danger. It is perfectly clear that [32]

> *Extra-mundane Space,* (if the material would be Finite in its Dimensions,) is not *imaginary,* but *Real.* Nor are void Spaces in the World, merely imaginary.

It is the same in respect to time: [33]

> Had God created the World *but This Moment,* it would not have been created at the Time it was created.

The denial of the possibility for God to give motion to the world is no more convincing: [34]

> And if God *has made* (or *can* make) Matter Finite in Dimensions, the *material Universe* must consequently be in its Nature *Moveable*; For nothing that is finite, is immoveable.

Leibniz's criticism of the concept of void space is, for Clarke, based on a complete misunderstanding of its nature and on misuse of metaphysical concepts: [35]

> *Space* void of Body, is the Property [attribute] of an *incorporeal* Substance. Space is not *Bounded* by *Bodies,* but exists equally *within* and *without* Bodies. Space is not

inclosed between Bodies; but Bodies, existing in unbounded Space, are, *themselves only*, terminated by their own Dimensions.

Void Space, is not an *Attribute without a Subject*, because, by *void Space*, we never mean *Space void* of *every thing*, but void of *Body* only. In All void *Space*, God is *certainly* present, and *possibly* many other Substances which are not Matter; being neither *Tangible*, nor Objects of any of *Our* Senses.

Space is not a *Substance*, but a *Property* [attribute]; And if it be a *Property* [attribute] of That which is necessary, it will consequently (as all *other* Properties [attributes] of That which is necessary must do), exist *more necessarily*, though it be not *itself* a Substance, than those *Substances Themselves* which are *not necessary*. Space is immense, and immutable, and eternal; and so also is *Duration*. Yet it does not at all from hence follow, that any thing is eternal *hors de Dieu*. For *Space* and *Duration* are not *hors de Dieu*, but are *caused by*, and are *immediate* and *necessary Consequences* of His Existence. And without them, his *Eternity* and *Ubiquity* (or *Omnipresence*) would be taken away.

Having thus established the ontological status of space as an attribute of God, Clarke proceeds to the demonstration that its attribution to God does not constitute a slur on His perfection: thus it does not make God divisible. Bodies are divisible, that is, can be broken up into parts,[36]

but infinite Space, though it may by Us be *partially apprehended*, that is, may in our Imagination be conceived as composed of *Parts*; yet Those *Parts* (*improperly* so called) being *essentially indiscerpible* [37] and *immoveable* from each other, and not *partable* without an express Contradiction

> in Terms, *Space* consequently is in itself *essentially One*, and *absolutely indivisible.*

It is this space which is a precondition of motion; and motion in the true and full sense of the word, is absolute motion, that is, motion in respect to this space, in which places, though perfectly similar, are nevertheless different. The reality of this motion proves, at the same time, the reality of absolute space: [38]

> It is largely insisted on by Sir *Isaac Newton* in his *Mathematical Principles* (Definit. 8) where, from the Consideration of the *Properties, Causes* and *Effects* of Motion, he shows the difference between *real Motion,* or a Bodie's being carried from one part of Space to another; and *relative Motion,* which is merely a change of the *Order* or *Situation* of Bodies with *respect to each other.*

The problem of time is exactly parallel to that of space: [39]

> It was no *impossibility* for God to make the World *sooner* or *later* than he did: Nor is it at all *impossible* for him to destroy it *sooner* or *later* than it shall actually be destroyed. As to the Notion of the *World's Eternity*; They who suppose *Matter* and *Space* to be the same, *must* indeed suppose the World to be not only *Infinite* and *Eternal,* but *necessarily so*; even as necessarily as *Space* and *Duration,* which depend not only on the *Will,* but on the *Existence* of God. But they who believe that God created Matter in what *Quantity,* and at what particular *Time,* and in what particular *Spaces* he *pleased,* are here under no difficulty. For the Wisdom of God may have *very good reasons* for creating *This World,* at *That* Particular Time he did.

Clarke's reasoning follows the well-trodden path: infinity implies necessity, and therefore: [40]

That *God Cannot limit the Quantity of Matter*, is an Assertion of too great consequence, to be admitted without *Proof*. If he cannot limit the *Duration* of it neither, then the material World is both infinite and eternal *necessarily* and *independently upon God.*

Thus we see it once more: the acceptance of absolute space as an attribute of God and as the universal container or receptacle of everything is the means — the only one — to avoid infinity, that is, self-sufficiency of matter, and to save the concept of creation: [41]

Space is the *Place* of *All Things*, and of All Ideas: Just as *Duration* is the *Duration* of *All Things*, and of *All Ideas*. . . . This has no Tendency to make God *the Soul* of the World.

Far from making God immersed in the world and thus, as Leibniz insinuates, dependent upon the world, the Newtonian conception is, according to Clarke, the only one that makes Him fully and truly independent of it; fully and truly free: [42]

There is no *Union* between *God* and the *World*. The *Mind of Man* might with greater propriety be stiled *The Soul of the Images of things which he perceives*, than God can be stiled *the Soul* of the World, to which he is *present* throughout, and *acts upon it* as he pleases, without being *acted upon by it.*

And it is just because of this independence of God from the world that [43]

. . . If *no Creatures* existed, yet the *Ubiquity* of God, and *Continuance of his Existence*, would make *Space* and *Duration* to be exactly the same as they are *Now.*

Finally, coming back to Leibniz's persistence in misunderstanding Newton's theory of attraction and in wanting to make it a miracle, Clarke (who pointed out that Leibniz's own theory of the "pre-established harmony" between the non-communicating and non-acting-upon-each-other mind and body has much more right to imply a perpetual miracle) explains,[44]

> That *One Body* should *attract* another *without any* intermediate *Means*, is indeed not a *Miracle*, but a *Contradiction*: For 'tis supposing something to *act* where it is *not*. But the *Means* by which Two Bodies attract each other, may be *invisible* and *intangible*, and of a different nature from *mechanism*; and yet, acting regularly and constantly, may well be called *natural*; being much less wonderful than *Animal-motion*, which yet is *never* called a *Miracle*.

Indeed, it is only from the point of view of the Cartesio-Leibnizian rigid dualism of mind and body, with its negation of all intermediate entities and consequent reduction of material nature to a pure, self-sustaining and self-perpetuating mechanism, that the intervention in nature of non-mechanical and therefore non-material agencies becomes a miracle. For Clarke, as for Henry More before him, this dualism is, of course, unacceptable. Matter does not constitute the whole of nature, but is only a part of it. Nature, therefore, includes both mechanical (*stricto sensu*) and non-mechanical forces and agencies, just as "natural" as the purely mechanical ones, material as well as immaterial entities which "fill" and pervade space and without which there would be no unity or structure in the world, or better to say, there would not be a world.

The world, of course, is not an organism, like the animal,

and possesses no "soul." Yet it can no more be reduced to pure mechanism than the animal, in spite of Descartes.

The vigorous (or, from Leibniz's point of view, obstinate) defense by Dr. Clarke of his (untenable) position; the assurance with which he not only accepted the (absurd and damaging) consequences deduced by Leibniz from his premises — the eternity of space — but even went beyond them by openly proclaiming that space (and time) were necessary and uncreated attributes of God; the lack of insight (or perfidy) with which he persisted in misinterpreting and misrepresenting Leibniz's principle of sufficient reason by identifying the supreme freedom of his supremely perfect God, unable to act except according to His supreme wisdom (that is, for the realization of the absolutely best universe unerringly recognized by Him among the infinite number of possible ones), with the fatality, necessity and passivity of a perfect mechanism, convinced Leibniz that he had to devote even more space and effort to the refutation of his adversary; and to the correction of the image that the latter presented of Leibniz's own views.

Thus the *fifth* (and last) paper addressed by Leibniz to the Princess of Wales became a lengthy treatise, the full analysis of which would lead us too far from our topic. It is, for us, sufficient to state that it starts with an admirable explanation of the difference between a *motive*, which inclines without compelling and thus preserves the spontaneity and the freedom of the subject, and a real cause, which necessarily produces its effect, and of the infinite distance that separates the moral — that is, free

— necessity of a fully motivated action from the unfree and passive necessity of a mechanism.

Freedom, indeed, for Leibniz, as for most philosophers, means doing what is good, or best, or what one ought to do, not simply doing what one wants to.[45] The laymen, alas — and Newton is no better than they — cannot make that distinction; they do not recognize freedom in the absolute determination of God's action. The laymen, and the theologians, therefore, accuse the philosophers of rejecting freedom in favor of necessity, and attribute to God actions utterly unworthy of Him. It is, however, evident that it is unreasonable to ask God to act in a purposeless irrational manner even if, strictly speaking, He is able — being all-powerful — to perform such an action. Thus, for instance: [46]

> Absolutely speaking, it appears that God *can* make the material Universe *finite* in Extension; but the contrary appears more agreeable to his Wisdom.

And it is, of course, even less " agreeable to his Wisdom " to move the world in a straight line — why, indeed, should God do such a meaningless thing? [47]

> And therefore the Fiction of a material finite Universe, moving forward in an infinite empty Space cannot be admitted. It is altogether unreasonable and *impracticable*. For, besides that there is *no real Space* out of the material Universe, such an Action would be without any Design in it: It would be working without doing any thing, *agendo nihil agere*. There would happen *no Change*, which could be observed by Any Person whatsoever. These are Imaginations of *Philosophers who have incomplete notions*, who make Space an absolute Reality.

Leibniz had already said it in his preceding paper, and even in stronger terms. Yet, in that paper he did not tell us *all* his reasons for rejecting this kind of motion. He did not mention precisely the most important one, namely that such a motion would be unobservable. It is perfectly clear that, if we accept the principle of observability, absolute motion, or at least absolute uniform motion in a straight line, which everybody agrees to be unobservable, will be ruled out as meaningless, and only relative motion will be acceptable. Yet in that case, the Newtonian formulation of the principle of inertia, stating that a body remains in its status of rest or uniform motion irrespective of what happens to others, and would remain in its status of motion or rest even if no other body existed, or if all of them were destroyed by God, will have to be rejected as meaningless and therefore impossible. But as it is only in such a case that the principle of inertia is fully valid, it is not only Newton's formulation of it, but the principle itself that becomes meaningless. These are rather far-reaching consequences of an innocent-looking principle, fully confirmed by the recent discussions about relativity, that are, as a matter of fact, an aftermath of the largely forgotten discussions of the XVIIth century.

Leibniz, of course, does not require that any and every motion be *actually* observed; yet, according to him, it must be possible to do so, and that for a rather surprising reason, a reason that shows us the depth of Leibniz's opposition to Newton, and the fidelity of Leibniz to old Aristotelian conceptions which modern science has been at such pains to reject and to reform: for Leibniz, indeed, motion is still conceived as a *change*, and not as a *status*: [48]

> . . . Motion does not indeed depend upon being *Observed*; but it does depend upon being *possible to be Observed.* There is no *Motion,* when there is no *Change* that can be *Observed.* And when there is no *Change that can be Observed,* there is *no Change at all.* The contrary Opinion is grounded upon the Supposition of a real absolute Space, which I have demonstratively confuted by the Principle of the want of a *sufficient Reason* of things.

The principle of observability confirms the relative character of motion and space. But relations — another far-reaching statement — have no "real", but only an "ideal", existence. Therefore,[49]

> since *Space* in it self is an *Ideal* thing, like *Time*; Space *out of the World* must needs be imaginary, as the *Schoolmen* themselves have acknowledged. The case is the Same with empty Space *within* the World; which I take also to be imaginary, for the reason before alleged.

The Schoolmen, to tell the truth, meant something quite different, and Leibniz knows it better than anyone: they conceived the world as finite and wanted to deny the existence of real space (and time) outside the world — Leibniz, on the contrary, denies the limitation of the universe. But in a sense he is right to appeal to them: for both time and space are intramundane and have no existence outside — or independently from — the created world. How, indeed, could time be something in itself, something real or even eternal? [50]

> It cannot be said, that *Duration* is Eternal; but that *Things,* which continue always, are Eternal. Whatever exists of Time and Duration, perishes continually: And how can a thing exist Eternally, which, (to speak exactly),

does never exist at all? For, how can a thing exist, whereof no Part does ever exist? Nothing of Time does ever exist, but Instants; and an Instant is not even it self a part of Time. Whoever considers These Observations, will easily apprehend that Time can only be an Ideal thing. And the Analogy between Time and Space, will easily make it appear that the one is as merely Ideal as the other.

Yet we must not unduly stress the parallelism between space and time in order not to be conduced to admit either the infinity of time, that is, the eternity of the world, or the possibility of a finite universe: [51]

. . . the World's having a Beginning, does not derogate from the Infinity of its Duration *a parte post*; but Bounds of the Universe would derogate from the Infinity of its Extension. And therefore it is more reasonable to admit a Beginning of the World, than to admit any Bounds of it; that the Character of its infinite Author, may be in Both Respects preserved.

However, those who have admitted the *Eternity* of the World, or, at least, (as some famous Divines have done) *the possibility* of its Eternity, did not, for all that, deny its dependence upon God; as the Author here lays to their Charge, without any Ground.

The Newtonians, of course, do not accept these Leibnizian " axioms " (and we have just seen that they have very good reasons for not doing so, as they overthrow the very foundations of their physics), and try to save absolute space by relating it to God. Leibniz, therefore, reminds us of his already formulated objections, which he repeats in the pious hope that, finally, he will succeed in convincing his opponent (or, at least, the Princess of

Wales) how utterly impossible it is to confer an absolute existence on void space.[52]

> I objected, that Space, taken for something real and absolute without Bodies, would be a thing eternal, impassible, and independent upon God. The Author endeavours to elude this Difficulty, by saying that Space is a property [attribute] of God.
>
> I objected further, that if Space be a property [attribute], and *infinite Space* be the *Immensity of God*; *finite Space* will be the *Extension* or *Mensurability* of something finite. And therefore the *Space* taken up by a *Body*, will be the *Extension of that Body*. Which is an absurdity; since a Body can change *Space*, but cannot leave its *Extension*.

Rather amusing to see Leibniz use against Clarke the same arguments that Henry More used against Descartes. But let us continue: [53]

> If infinite *Space* is God's *Immensity*, infinite *Time* will be God's *Eternity*; and therefore we must say, that what is in Space, is in God's Immensity, and consequently in his Essence; and that what is in Time, is also in the Essence of God. *Strange* Expressions; which plainly show, that the Author makes a wrong use of Terms.

Assuredly, at least if we follow the traditional scholastic conceptions. But the Newtonians, as we know, reinterpret these terms and expressly identify God's immensity with infinite extension and God's eternity with infinite duration. They will therefore acknowledge that everything is *in* God, without being obliged to put everything in his *essence*. But Leibniz insists: [54]

> I shall give another Instance of This. God's Immensity makes him actually present in all Spaces. But now if God

is *in* Space, how can it be said that Space is *in* God, or that it is a Property [attribute] of God? We have often heard, that a Property [attribute] is in its Subject; but we never heard, that a Subject is in its Property [attribute]. In Like manner, God exists *in* all Time. How then can Time be *in* God; and how can it be a Property [attribute] of God? These are perpetual *Alloglossies*.

Once more, the Newtonians would object that the preposition *in* is obviously taken in two different meanings, and that nobody has ever interpreted the attribute being *in* the substance as a spatial relation; that, moreover, they only draw a correct conclusion from God's omnipresence, which everybody admits, and God's simplicity, which everybody admits also, by refusing to recognize, in God, a separation between His substance and His power and asserting therefore His substantial presence everywhere. They would deny Leibniz's contention that [55]

It appears that the Author confounds Immensity or the *Extension of Things*, with the *Space* according to which that Extension is taken. Infinite Space, is not the Immensity of God; Finite Space, is not the Extension of Bodies: As Time is not their Duration. Things keep their Extension; but they do not always keep their Space. Every Thing has its own Extension, its own Duration; but it has not its own Time, and does not keep its own Space.

Of course not. But for the Newtonians, it means precisely that time and space do not belong to things, nor are relations based upon the existence of things, but belong to God as a framework in which things and events have and take place. Leibniz knows it, of course, but he cannot admit this conception: [56]

Space is not the Place of all Things; for it is not the Place of *God.* Otherwise there would be a thing co-eternal with God, and independent upon him; nay, he himself would depend upon *it,* if he has need of *Place.*

If the reality of Space and Time, is necessary to the Immensity and Eternity of God, if God must be in Space; if being in Space is a Property [attribute] of God; he will, in some measure, depend upon Time and Space, and stand in need of them. For I have already prevented That Subterfuge, that Space and Time are *Properties* [attributes] of God.

Still, Leibniz knows that his own position implies difficulties (they are not proper to it, but are those of the whole scholastic tradition): if space and time are only innerworldly entities, and did not exist before Creation, must we not assume that the creation of the world brought about change in God; and that, before it, He was neither immense nor omnipresent? is not, therefore, God, in his own conception, dependent upon creatures? Leibniz writes then: [57]

'Tis true, the Immensity and Eternity of God would subsist, though there were no Creatures; but those Attributes would have no dependence either on *Times* or *Places.* If there were no Creatures, there would be neither *Time* nor *Place,* and consequently no actual *Space.* The Immensity of God is independent upon *Space,* as Eternity is independent upon *Time.* These attributes signify only, that God would be present and co-existent with all the Things that should exist.

A perfect answer. . . . Alas, the Newtonian will not accept it, and will persist in his affirmation that though, of course, God cannot be co-present with things that do not exist, their existence or non-existence does not make him more, or less, present in those places where these things, once created, will co-exist with him.

Having dealt with the general problem of space and time, Leibniz passes to the re-examination of the particular problem of attraction. Dr. Clarke's explanation did not satisfy him; quite the contrary. A miracle is not defined by its being an exceptional and rare happening: a miracle is defined by the very nature of the event. Something that cannot be explained *naturally*, that is, something that cannot result from the interplay of *natural* forces, that is, forces derived from the nature of things, is and remains a miracle. Now the nature of things does not admit action at a distance. Attraction therefore would be a miracle, though a perpetual one. Moreover, according to Leibniz, the suggestion made by Dr. Clarke to explain it by the action of non-mechanical, " spiritual " forces, is even worse; this, indeed, would mean going back behind Descartes, renouncing science for magic. Once more we see expressed in this debate the radical opposition of two conflicting views of nature, and of science: Leibniz can accept neither the Newtonian conception of the insufficiency of the material nature nor the provisional positivism of his conception of " mathematical philosophy ": [58]

> I objected, that an *Attraction*, properly so called, or in the *Scholastic* Sense, would be an Operation at a Distance, without any *Means* intervening. The Author answers here, that an *attraction* without any *means* intervening would be indeed a Contradiction. Very well! But then what does he mean, when he will have the Sun to attract the Globe of

the Earth through an empty Space? It is God himself that performs it? But this would be a *Miracle*, if ever there was any. This would surely exceed the Powers of Creatures.

Or, are perhaps some immaterial Substances, or some spiritual Rays, or some Accident without a Substance, or some Kind of *Species Intentionalis*, or some other *I know not what*, the Means by which this is pretended to be performed? Of which sort of things, the Author seems to have still a good stock in his Head, without explaining himself sufficiently?

That Means of communication (says he) is invisible, intangible, not Mechanical. He might as well have added, inexplicable, unintelligible, precarious, groundless, and unexampled.

If the Means, which causes an *Attraction* properly so called, be constant, and at the same time inexplicable by the Powers of Creatures, and yet be true; it must be a perpetual *Miracle*: And if it is not miraculous, it is false. 'Tis a Chimerical Thing, a Scholastic *occult quality*.

The Case would be the same, as in a Body going round without receding in the Tangent, though nothing that can be explained, hindered it from receding. Which is an Instance I have already alleged; and the Author has not thought fit to answer it, because it shows too clearly the difference between what is truely *Natural* on the one side, and a *chimerical occult Quality* of the Schools on the other.

Once more Dr. Clarke replied. He was, needless to say, not convinced. Leibniz's subtle distinctions did not succeed in hiding the brute fact that his God was subjected to a strict and unescapable determinism. He lacked not only the true freedom that belongs to a spiritual being but even the spontaneity (Leibniz, moreover, seemed to Clarke to confound the two) belonging to an animal one: He was no more than a pure mechanism enchained by an absolute necessity. If Dr. Clarke had the gift of foreseeing things, he would say: a mere calculating machine!

Leibniz's renewed attack on Newton's conceptions of time, space and motion is not more successful.[59]

> It is affirmed, that Motion necessarily implies a *Relative Change of Situation in one Body, with regard to other Bodies*; And yet no way is shown to avoid this absurd Consequence, that then the *Mobility* of *one Body* depends on the *Existence of other Bodies*; and that any *single Body* existing *Alone*, would be *incapable of Motion*; or that the Parts of a *circulating* Body (suppose the Sun) would lose the *vis centrifuga* arising from their circular Motion, if all the extrinsick Matter around them were annihilated, 'tis affirmed that the *Infinity of Matter* is an Effect of the *Will of God*.

And yet, if it were true that — as taught by Descartes — a finite universe is contradictory, is it not clear that, in this case, God neither is, nor was, able to limit the quantity of matter and therefore did not create, and can not destroy it? Indeed,[60]

> if the *Material Universe* CAN *possibly*, by the Will of God, be *finite* and *Moveable*: (which this learned Author here finds himself necessitated to *grant*, though he perpetually treats it as an *impossible* supposition;) then *Space* (in which That Motion is performed) is manifestly *independent* upon *Matter*. But if, on the contrary, the *material Universe Cannot* be *finite* and *moveable* and *Space cannot* be *independent* upon *Matter*; then (I say) it follows evidently, that God neither *Can* nor *ever Could* set Bounds to Matter; and consequently the *material Universe* must be not only

boundless, but *eternal* also, both *a parte ante* and *a parte post necessarily* and *independent of the Will of God.*

As for the relation between space, body and God, Clarke restates his position with perfect clarity: [61]

The space occupied by a body is not the extension of that body; but the extended body exists in this space.

There is no bounded space; but our imagination considers in the space, which has no limits and cannot have any, such a part, or such a quantity that it judges convenient to consider.

Space is not the affection of one or several bodies, nor that of any bounded thing, and it does not pass from one subject to another, but it is always, and without variation, the immensity of an immense being, which never ceases to be the same.

Bounded spaces are not properties of bounded substances; they are only parts of the infinite space in which the bounded substances exist.

If matter were infinite, infinite space would no more be a property of this infinite body than finite spaces are properties of finite bodies. But, in this case, infinite matter would be in infinite space as finite bodies are in it now.

Immensity, as well as Eternity, is essential to God. The *Parts* of *Immensity,* (being totally of a different Kind from *corporeal, partable, separable, divisible, moveable* Parts, which are the ground of *Corruptibility*), do no more hinder *Immensity* from being essentially *One,* than the *Parts* of *Duration* hinder *Eternity* from being essentially One.

God himself is not subjected to any change by the diversity and the change of things that are in him, and which in him have life, motion and being.

This *strange* Doctrine is the express Assertion of *St. Paul,* as well as the plain Voice of *Nature and Reason.*

> God is not in space or in time; but his existence is the cause of space and time. And when we say, in conformity with the language of the vulgar, that God exists in all the spaces and in all the times,
>
> These Words mean only that he is *Omnipresent* and *Eternal*, that is, that *Boundless Space* and *Time* are necessary *Consequences* of his Existence; and not, that Space and Time are Beings distinct from him, and IN which he exists.

Moreover,[62]

> to say that *Immensity* does not signify *Boundless Space*, and that *Eternity* does not signify *Duration* or *Time without Beginning and End*, is (I think) affirming that *Words* have no *meaning*.

As for the criticism of attraction, Clarke, of course, maintains his point of view: miracles are rare and meaningful events produced by God for definite reasons; a perpetual miracle is a contradiction in terms; and if not, then the *pre-established Harmony* of Leibniz is a much greater one. Moreover — Clarke is rather astonished that Leibniz does not understand this — in Newtonian science or *mathematical* philosophy, attraction (whatever be its ultimate physical or metaphysical explanation) appears only as a phenomenon, as a general fact and as a mathematical expression. Therefore,[63]

> it is very unreasonable to call *Attraction* a *Miracle* and an unphilosophical Term; after it has been so often distinctly declared, that by That Term we do not mean to express the Cause of Bodies tending *towards each other*, but barely the *Effect*, or the *Phaenomenon it self*, and the *Laws* or *Proportions of that Tendency*, discovered by *Experience*

which clearly shows

> that the Sun *attracts* the Earth, through the intermediate void Space; that is that the Earth and Sun *gravitate* towards each other, or *tend* (whatever be the Cause of that Tendency) towards each other, with a Force, which is in a direct *proportion* of their *Masses*, or *Magnitudes and Densities together*, and in an inverse duplicate proportion of their Distances.

But, of course, there is much more behind this Leibnizian opposition to attraction than a mere unwillingness to adopt the point of view of "mathematical" philosophy with its admission into the body of science of incomprehensible and inexplicable "facts" imposed upon us by empiricism: what Leibniz really aims at is the self-sufficiency of the world-mechanism, and there is very little doubt that the law of conservation of the *vis viva* achieves it in a still better way than the Cartesian law of conservation of motion.

The Newtonian world — a clock running down — requires a constant renewal by God of its energetic endowment; the Leibnizian one, by its very perfection, rules out any intervention of God into its perpetual motion. Thus it is not surprising that for Dr. Clarke the fight for void space, hard atoms and absolute motion becomes a fight for God's Lordship and presence, and that he asks Leibniz why [64]

> . . . so great Concern should be shown, to exclude God's *actual* Government of the World, and to allow his Providence to *act* no further than barely in *concurring* (as the Phrase is) to let *all Things* do only what they would do *of themselves of mere Mechanism.*

XII. Conclusion:

THE DIVINE ARTIFEX AND THE *DIEU FAINÉANT*

Why, indeed? Leibniz, who was much more interested in morals than in physics and in man than in the cosmos, could have answered that it was the only means to avoid making God responsible for the actual management, or mismanagement, of this our world. God just did not do what He wanted, or would like to do. There were laws, and rules, that He could neither change nor tamper with. Things had natures that He could not modify. He had made a perfect mechanism in the working of which He could not interfere. Could not and should not, as this world was the best of all the possible worlds that He could create. God, therefore, was blameless for the evils that He could not prevent or amend. After all, this world was only the best *possible* world, not a perfectly good one; that was *not* possible.

Leibniz might have said this in reply to Clarke. But he did not read Clarke's *fifth* reply. He died before he received it. Thus their fight, a fight in which both sides fought *pro majore Dei gloria*, ended as abruptly as it started. The outcome of the Homeric struggle was not conclusive; neither side, as we have seen, budged an inch. Yet, in the decades that followed, Newtonian science and

Newtonian philosophy gained more and more ground, gradually overcoming the resistance of the Cartesians and the Leibnizians who, though opposing each other on many points, made a common front against the common foe.

At the end of the century Newton's victory was complete. The Newtonian God reigned supreme in the infinite void of absolute space in which the force of universal attraction linked together the atomically structured bodies of the immense universe and made them move around in accordance with strict mathematical laws.

Yet it can be argued that this victory was a Pyrrhic one, and that the price paid for it was disastrously high. Thus, for instance, the force of attraction which, for Newton, was a proof of the insufficiency of pure mechanism, a demonstration of the existence of higher, non-mechanical powers, the manifestation of God's presence and action in the world, ceased to play this role, and became a purely natural force, a property of matter, that enriched mechanism instead of supplanting it. As Dr. Cheyne explained quite reasonably, attraction was assuredly not an essential property of body, but why should not God have endowed matter with unessential properties? Or, as Henry More and Roger Cotes — and later, Voltaire — pointed out, since we possess no knowledge of the substances of things, and know nothing about the link that connects property with substance, even in the cases of hardness or impenetrability, we cannot deny that attraction belongs to matter just because we do not understand how it works.

As for the dimensions of the material universe which Newtonians at first had opposed to the actual infinity of absolute space, the relentless pressure of the principles of plenitude and sufficient reason, by which Leibniz managed to infect his successful rivals, made it co-extensive with space itself. God, even the Newtonian one, could obviously not limit His creative action and treat a certain part of infinite homogeneous space — though able to distinguish it from the rest — in a way so utterly different from the others. Thus the material universe, in spite of filling only an exceedingly small part of the infinite void, became just as infinite as this. The same reasoning which prevented God from limiting His creative action in respect to space could, just as well, be applied to time. An infinite, immutable and *sempiternal* God could not be conceived as behaving in a different manner at different times, and as limiting His creative action to a small stretch of it. Moreover, an infinite universe existing only for a limited duration seems illogical. Thus the created world became infinite both in Space and in Time. But an infinite and eternal world, as Clarke had so strongly objected to Leibniz, can hardly admit creation. It does not need it; it exists by virtue of this very infinity.

Furthermore, the gradual dissolution of traditional ontology under the impact of the new philosophy undermined the validity of the inference from the attribute to its supporting substance. Space, consequently, lost progressively its attributive or substantial character; from the ultimate stuff which the world was made of (the substantial space of Descartes) or the attribute of God, the frame of his presence and action (the space of Newton), it became more and more the void of the atomists, neither substance nor accident, the infinite, uncreated nothingness, the frame of the absence of all being; consequently also of God's.

Last but not least, the world-clock made by the Divine Artifex was much better than Newton had thought it to be. Every progress of Newtonian science brought new proofs for Leibniz's contention: the moving force of the universe, its *vis viva,* did not decrease; the world-clock needed neither rewinding, nor mending.

The Divine Artifex had therefore less and less to do in the world. He did not even need to conserve it, as the world, more and more, became able to dispense with this service.

Thus the mighty, energetic God of Newton who actually " ran " the universe according to His free will and decision, became, in quick succession, a conservative power, an *intelligentia supra-mundana,* a " Dieu fainéant."

Laplace who, a hundred years after Newton, brought the New Cosmology to its final perfection, told Napoleon, who asked him about the role of God in his *System of the World:* " Sire, je n'ai pas eu besoin de cette hypothèse." But it was not Laplace's *System,* it was the world described in it that no longer needed the hypothesis God.

The infinite Universe of the New Cosmology, infinite in Duration as well as in Extension, in which eternal matter in accordance with eternal and necessary laws moves endlessly and aimlessly in eternal space, inherited all the ontological attributes of Divinity. Yet only those — all the others the departed God took away with Him.

Notes

INTRODUCTION AND CHAPTER I

1. Cf. A. N. Whitehead, *Science and the modern world*, New York, 1925; E. A. Burtt, *The metaphysical foundations of modern physical science*, New York, 1926; J. H. Randall, *The making of the modern mind*, Boston, 1926; Arthur O. Lovejoy's classical *Great chain of being*, Cambridge, Mass., 1936, and my own *Études Galiléennes*, Paris, 1939.
2. The cosmos conception is only practically, that is, historically, linked together with the geocentric world-view. Yet it can be completely divorced from the latter as, for example, by Kepler.
3. The full story of the transformation of the space conception from the Middle Ages to modern times should include the history of the Platonic and Neoplatonic revival from the Florentine Academy to the Cambridge Platonists as well as that of the atomic conceptions of matter and the discussions about the vacuum following the experiments of Galileo, Torricelli and Pascal. But this would double the volume of this work and, besides, distract us somewhat from the very definite and precise line of development which we are following here. Moreover, for some of these problems we can refer our readers to the classical books of Kurd Lasswitz, *Geschichte des Atomistik*, 2 vols., Hamburg und Berlin, 1890, and Ernst Cassirer, *Das Erkenntnisproblem in der Philosophie und Wissenschaft der neuen Zeit*, 2 vols., Berlin, 1911, as well as to the recent works of Cornelis de Waard, *L'expérience barométrique, ses antécédents et ses explications*, Thouars, 1936, and Miss Marie Boas, "Establishment of the mechanical philosophy," *Osiris*, vol. x, 1952. See now Max Jammer, *Concepts of space*, Harvard Univ. Press, Cambridge, Mass., 1954, and Markus Fierz, "Ueber den Ursprung und Bedeutung von' Newtons Lehre vom absoluten Raum," *Gesnerus*, vol. xi, fasc. 3/4, 1954, especially for the space conceptions of Telesio Pattrizzi and Campanella.
4. On the Greek conceptions of the universe cf. Pierre Duhem, *Le système du monde*, vol. i and ii, Paris, 1913, 1914; R. Mondolfo, *L'infinito nel pensiero dei Greci*, Firenze, 1934, and Charles Mugler, *Devenir cyclique et la pluralité des mondes*, Paris, 1953.

5. The MS of *De rerum natura* was discovered in 1417. On its reception and influence cf. J. H. Sandys, *History of classical scholarship*, Cambridge, 1908, and G. Hadzitz, *Lucretius and his influence*, New York, 1935.
6. The first Latin translation of Diogenes Laertius' *De vita et moribus philosophorum* by Ambrosius Civenius appeared in Venice in 1475 and was immediately reprinted in Nürnberg in 1476 and 1479.
7. The atomism of the ancients, at least in the aspect presented to us by Epicurus and Lucretius — it may be that it was different with Democritus, but we know very little about Democritus — was not a scientific theory, and though some of its precepts, as for instance, that which enjoins us to explain the celestial phenomena on the pattern of the terrestrial ones, *seem* to lead to the unification of the world achieved by modern science, it has never been able to yield a foundation for development of a physics; not even in modern times: indeed, its revival by Gassendi remained perfectly sterile. The explanation of this sterility lies, in my opinion, in the extreme sensualism of the Epicurean tradition; it is only when this sensualism was rejected by the founders of modern science and replaced by a mathematical approach to nature that atomism — in the works of Galileo, R. Boyle, Newton, etc. — became a scientifically valid conception, and Lucretius and Epicurus appeared as forerunners of modern science. It is possible, of course, and even probable, that, in linking mathematism with atomism, modern science revived the deepest intuitions and intentions of Democritus.
8. Cf. René Descartes, "Lettre à Chanut," June 6, 1647, *Oeuvres*, ed. Adam Tannery, vol. v, p. 50 sq., Paris, 1903.
9. Nicholas of Cusa (Nicholas Krebs or Chrypffs) was born in 1401 in Cues (or Cusa) on the Moselle. He studied law and mathematics in Padua, then theology in Cologne. As archdeacon of Liège he was a member of the Council of Basel (1437), was sent to Constantinople to bring about a union of the Eastern and Western churches, and to Germany as papal legate (1440). In 1448 he was raised by Pope Nicholas V to the cardinalate, and in 1450 he was appointed Bishop of Britten. He died August 11, 1464. On Nicholas of Cusa cf. Edmond Vansteenberghe, *Le Cardinal Nicolas de Cues*, Paris, 1920; Henry Bett, *Nicolas of Cusa*, London, 1932; Maurice de Gandillac, *La philosophie de Nicolas de Cues*, Paris, 1941.
10. Cf. Ernst Hoffmann, *Das Universum von Nikolas von Cues*, especially the *Textbeilage* by Raymond Klibansky, pp. 41 sq., which gives the text of Nicholas of Cusa in a critical edition as well as the bibliography of the problem. The booklet of E. Hoffmann appeared as "Cusanus

Studien, I" in the *Sitzungsberichte der Heidelberger Akademie der Wissenschaften, Philosophisch-Historische Klasse,* Jahrgang 1929/1930, 3. Abhandlung, Heidelberg, 1930.

11. Cf. *De docta ignorantia,* l. II, cap. ii, p. 99. I am following the text of the latest, critical, edition of the works of Nicholas of Cusa by E. Hoffmann-R. Klibansky (*Opera omnia, Jussu et auctoritate Academiae litterarum Heidelbergensii ad codicum fidem edita,* vol. I, Lipsiae, 1932). There is, now, an English translation of the *De docta ignorantia* by Fr. Germain Heron: *Of learned ignorance* by Nicholas Cusanus, London, 1954. I have, nevertheless, preferred to give my own translation of the texts I am quoting.
12. *Ibid.,* p. 99 sq.
13. *Ibid.,* p. 100.
14. *Ibid.,* p. 100 sq. It is to be remembered, however, that the conception of the relativity of motion, at least in the sense of the necessity to relate motion to a resting reference-point (or body) is nothing new and can already be found in Aristotle; cf. P. Duhem, *Le mouvement absolu et le mouvement relatif,* Montlignon, 1909; the optical relativity of motion is studied at length by Witello (cf. *Opticae libri decem,* p. 167, Basilae, 1572) and, even more extensively, by Nicole Oresme, (cf. *Le livre du ciel et de la terre,* ed. by A. D. Meuret and A. J. Denomy, C. S. B., pp. 271 sq., Toronto, 1943).
15. *Ibid.,* p. 102.
16. *Ibid.,* p. 102 sq.
17. *De docta ignorantia,* l. II, cap. 12, p. 103.
18. Cf. the famous passage of Virgil, *Provehimur portu terraeque urbesque recedunt,* quoted by Copernicus.
19. This famous saying which describes God as a *sphaera cuius centrum ubique, circumferentia nullibi* appears for the first time in this form in the pseudo-Hermetic *Book of the XXIV philosophers,* an anonymous compilation of the XIIth century; cf. Clemens Baemker, *Das pseudo-hermetische Buch der XXIV Meister* (Beiträge zur Geschichte der Philosophie und Theologie des Mittelalters, fasc. xxv), Münster, 1928; Dietrich Mahnke, *Unendliche Sphaere und Allmittelpunct,* Halle/Saale, 1937. In this *Book of the XXIV philosophers,* the above-mentioned formula forms the proposition II.
20. He is, however, referred to by Giovanni Francesco Pico in his *Examen doctae vanitatis gentium* (*Opera,* t. II, p. 773, Basileae, 1573) and Celio Calcagnini in his *Quod coelum stet, terra moveatur, vel de perenni motu terrae* (*Opera aliquot,* p. 395, Basileae, 1544); cf. R. Klibansky, *op. cit.,* p. 41.

21. Cf. L. A. Birkenmajer, *Mikolaj Kopernik,* vol. I, p. 248, Cracow, 1900. Birkenmajer denies any influence of Nicholas of Cusa on Copernicus. On the medieval "forerunners" of Copernicus cf. G. McColley, "The theory of the diurnal rotation of the earth," *Isis,* XXVI, 1937.
22. *De docta ignorantia,* II, 12, p. 104.
23. Nicholas of Cusa's conception could be treated as an anticipation of that of Sir William Herschell; and even of more modern ones.
24. *De docta ignorantia,* II, 12, p. 104.
25. *Ibid.,* p. 105.
26. *Ibid.,* p. 107. Once more, one could see in this conception of Nicholas of Cusa a prefiguration of the theory of the mutual attraction of the heavenly bodies.
27. *Ibid.,* p. 107.
28. *Ibid.,* p. 108 sq.
29. Marcellus Stellatus Palingenius, whose true name was Pier Angelo Manzoli, born at La Stellata some time between 1500 and 1503, wrote, under the title of *Zodiacus vitae,* a didactical poem, which was printed in Venice (probably) in 1534, rapidly became extremely popular, especially among Protestants, and was even translated into English, French and German. The English translation (*Zodiake of life*) by Barnaby Goodge, appeared in 1560 (the first three books), and in 1565 the entire poem was printed. It seems that Palingenius was at a certain time suspected of heresy, but it was only 15 years after his death (he died in 1543) that, in 1558, the *Zodiacus vitae* was put on the *Index librorum prohibitorum.* Under Pope Paul II his bones were disinterred and burnt; cf. F. W. Watson, *The Zodiacus Vitae of Marcellus Palingenius Stellatus: An old school book,* London, 1908 and F. R. Johnson, *Astronomical thought in Renaissance England,* pp. 145 sq., Baltimore, 1937.
30. *Zodiacus vitae,* l. VII, *Libra,* ll. 497-99; Engl. transl., p. 118; cf. A. O. Lovejoy, *The great chain of being,* pp. 115 sq., Cambridge, Mass., 1936; F. R. Johnson, *op. cit.,* pp. 147 sq.
31. *Zodiacus vitae,* l. IX, *Aquarius,* ll. 601-3 (transl., p. 218).
32. *Ibid.,* l. XI, *Aquarius,* ll. 612-616 (transl., p. 218).
33. A. O. Lovejoy, *The great chain of being,* p. 52 and *passim.*
34. *Zodiacus vitae,* l. XII, *Pisces,* ll. 20-35 (transl., p. 228).
35. *Ibid.,* ll. 71-85 (transl., p. 229). The world-view of Palingenius is beautifully expressed by Edmund Spenser in his *Hymn of heavenly beauty* (quoted by E. M. W. Tillyard, *The Elizabethan world picture,* p. 45, London, 1943):

    Far above these heavens which here we see,

Be others far exceeding these in light,
Not bounded, not corrupt, as these same be,
But infinite in largeness and in height,
Unmoving, incorrupt and spotless bright
That need no sun t'illuminate their spheres
But their own native light far passing theirs
And as these heavens still by degree arise
Untill they come to their first mover's bound,
That in his mighty compass doth comprise
And carry all the rest with him around;
So those likewise do by degree redound
And rise more fair till they at last arrive
To the most fair, whereto they all do strive.

CHAPTER II

1. In the technical sense of the word, Copernicus is a Ptolemean.
2. Cf. Joachim Rheticus, *Narratio prima.* I am quoting the excellent translation of E. Rosen in his *Three Copernican treatises,* p. 147, New York, 1939.
3. F. R. Johnson, *Astronomical thought in Renaissance England,* pp. 245-49, Baltimore, 1937; cf. A. O. Lovejoy, *op. cit.*, pp. 109 sq.
4. John Donne, *Anatomy of the world,* First Anniversary (1611) ed., Nonesuch Press, p. 202. The disastrous effects of the seventeenth century's spiritual revolution have recently been studied with great care and some nostalgic regret by a number of scholars; cf. *inter alia,* E. M. W. Tillyard, *The Elizabethan world picture,* London, 1943; Victor Harris, *All coherence gone,* Chicago, 1949; Miss Marjorie H. Nicolson, *The breaking of the circle,* Evanston, Ill., 1950; S. L. Bethell, *The cultural revolution of the XVIIth century,* London, 1951. For a non-nostalgic treatment cf. A. O. Lovejoy, *The great chain of being,* and Basil Willey, *The seventeenth century background,* Cambridge, 1934.
5. Nicolaus Copernicus, *De revolutionibus orbium coelestium,* l. I, cap. VIII.
6. According to the mediaeval conception the central position of the earth is the lowest possible; only Hell is " lower " than our earthly abode.
7. For the pre-modern, that is, pre-telescopic astronomy, fixed stars possess a visible and even measurable diameter. Since, on the other hand, they are rather far away from us and in the Copernican conception even exceedingly far (cf. *infra,* pp. 92-9), their real dimensions must be extremely large.

8. Cf. Grant McColley, "The seventeenth century doctrine of a plurality of worlds," *Annals of Science*, I, 1936, and "Copernicus and the infinite universe," *Popular Astronomy*, XLIV, 1936; cf. Francis R. Johnson, *op. cit.*, pp. 107 sq.
9. Nicolaus Copernicus, *De revolutionibus orbium coelestium*, l. I, cap. I.
10. *Ibid.*, l. I, cap. VIII.
11. *Ibid.*, l. I, cap. X.
12. A. O. Lovejoy, *op. cit.*, pp. 99 sq.
13. Cf. Sir Walter Raleigh, *The historie of the world*, London, 1652, pp. 93 sq.; cf. Bethell, *op. cit.*, pp. 46 sq.
14. Cf. *infra*, p. 94.
15. Giordano Bruno understands them as teaching the infinity of the universe. I have already examined the case of Nicholas of Cusa; as for Lucretius, he asserts, indeed, the infinity of space an[illegible] *worlds*, but maintains the finiteness of our visible world and [illegible] of a limiting heavenly sphere, outside of which, but inaccessible to our perception, there are other identical or analogous "worlds." Anachronistically we could consider his conception as prefigurating the modern conception of island-universes dispersed in an infinite space, though with a very important difference: the Lucretian worlds are closed and not connected with each other.
16. Cf. Francis R. Johnson and Sanford V. Larkey, "Thomas Digges, the Copernican system and the idea of the infinity of the universe," *The Huntington Library Bulletin*, n. 5 (1934), and Francis R. Johnson, *op. cit.*, pp. 164 sq.; cf. also A. O. Lovejoy, *op. cit.*, p. 116.
18. *A Perfit Description*, sigs N 3-N 4; cf. Johnson-Larkey, pp. 88 sq.; Johnson, pp. 165-167.
19. A. O. Lovejoy, *op. cit.*, p. 116. Giordano Bruno was born in Nola (near Naples) in 1548, became a Dominican in 1566, but, ten years later in 1576, on account of some rather heretical views held by him on transsubstantiation and the Immaculate Conception, had to leave both the order and Italy. In 1579 he came to Geneva (where he could not stay), then to Toulouse, and to Paris (1581) where he lectured on the logical system of Raymundus Lullus (and wrote some philosophical works, i. e. *De umbris idearum* and a satiric comedy, *Il Candelajo*); in 1583 he went to England where he lectured and published some of his best works, such as *La Cena de le Ceneri, De la causa, principio et uno* and *De l'infinito universo e mondi*. From 1585 to 1592 Bruno wandered in Europe (Paris, Marburg, Wittenberg, Prague, Helmstadt, Zürich), publishing the *De immenso et innumerabilibus* in 1591. Finally, in 1592 he accepted an invitation to Venice. Denounced and arrested by the

Inquisition (in 1593), he was brought to Rome, where he remained imprisoned for seven years, until he was excommunicated and burnt at the stake on February 17, 1600. Cf. Dorothea Waley Singer *Giordano Bruno, his life and thought,* New York, 1950.

20. Written in 1584.
21. Cf. my *Études Galiléenes,* III, p. ii sq., and "Galileo and the scientific revolution of the XVIIth century," *The Philosophical Review,* 1943.
22. Giordano Bruno, *La Cena de le Ceneri,* dial. terzo, *Opere Italiane,* ed. G. Gentile, vol. I, p. 73, Bari, 1907.
23. *Ibid.,* pp. 73 sq.
24. The *De l'infinito universo e mondi* was written in 1584; the *De immenso et innumerabilibus,* or to quote the full title, *De innumerabilibus, immenso et infigurabili: sive de universo et mundis libri octo,* in 1591. I shall base my exposition on the *De l'infinito universo e mondi* and quote it in the excellent recent translation of Mrs. Dorothea Waley Singer, adjoined to her *Giordano Bruno, his life and work,* New York, 1950. I shall give the reference first to the edition of Gentile (*Opere Italiane,* vol. I); then to Mrs. Singer's translation.
25. Bruno's space is a void; but this void is nowhere really void; it is everywhere full of being. A vacuum with nothing filling it would mean a limitation of God's creative action and, moreover, a sin against the principle of sufficient reason which forbids God to treat any part of space in a manner different from any other.
26. *De l'inf. univ. e mondi,* p. 309 sq., transl., p. 280; cf. *De immenso . . . Opera latina,* vol. I, part I, p. 259.
27. A. O. Lovejoy, *op. cit.,* p. 119.
28. *De l'inf. universo,* dedic. epistle, p. 275 (transl., p. 246).
29. The famous phrase "le silence éternel de ces espaces infinis m'effraye" does not express Pascal's own feeling — as is usually assumed by Pascal's historians — but that of the atheistic "libertin."
30. *De l'inf. universo,* p. 274 (transl., p. 245).
31. *De l'inf. universo,* p. 280 (transl., p. 250); cf. *De immenso,* I, 4, *Opera,* I, I, p. 214.
32. *Ibid.,* p. 281 (transl., p. 251).
33. This very famous argument against the finitude of the universe — or of space — is a good example of the continuity of philosophical tradition and discussion. Giordano Bruno probably borrows it from Lucretius (*De rerum natura,* l. I, v. 968 sq.), but it was already widely used in the discussions of the XIII-XIVth centuries about the plurality of the worlds and the possibility of the void (cf. my paper quoted in chap. III, 40) and will be used by Henry More (cf. *infra,* p. 139) and even by Locke

(cf. *An essay on human understanding,* l. II, §§13, 21). According to the *Commentaire exégétique et critique* of A. Ernout and L. Robin to their edition of the *De rerum natura* (p. 180 sq., Paris, 1925), the argument originates with Architas and is used by Endemios in his *Physics* (cf. H. Diels, *Fragmente der Vorsocratiker,* c. XXXV, A 24, Berlin, 1912). What is more important, it is to be found in Cicero, *De natura deorum,* I, 20, 54; cf. Cyril Bailey, Lucretius, *De rerum natura,* vol. II, pp. 958 sq., Oxford, 1947.

34. *De l'inf. universo,* p. 282 (transl., p. 253).
35. *Ibid.,* p. 283 (transl., p. 254); cf. *Acrotismus Camoeracensis, Opera,* I, I, pp. 133, 134, 140.
36. Cf. *Acrotismus Camoeracensis,* p. 175.
37. *De l'inf. univ.,* p. 286 (transl., p. 256).
38. *Ibid.,* p. 289 (transl., p. 259).
39. *Ibid.,* p. 334 (transl., p. 304); cf. *De immenso, Opera,* I, I, p. 218.
40. *Ibid.,* p. 335 (transl., p. 304); cf. *De immenso, Opera,* I, I, p. 290; I, II, p. 66.
41. *Ibid.,* p. 336 (transl., p. 305); cf. *De immenso,* I, II, p. 121.
42. *Ibid.,* p. 336 (transl., p. 305).
43. *Ibid.,* p. 286 (transl., p. 257).
44. *Ibid.,* p. 289 (transl., p. 260).
45. As a scientist he was, sometimes, far behind it.
46. Cf. F. R. Johnson, *Astronomical thought in Renaissance England,* p. 216.
47. *G. Guillielmi Gilberti Colcestrensis, medici Londinensis, De magnete, magnetisque corporibus, et de magno magnete tellure physiologia nova,* c. VI, cap. III; pp. 215 sq., London, 1600; Gilbert's work was translated by P. Fleury Mottelay in 1892 and by Sylvanus P. Thompson in 1900. The Mottelay translation was reprinted in 1941 as one of " The Classics of the St. John's Program " under the title: *William Gilbert of Colchester, physician of London, On the load stone and magnetic bodies and on the great magnet the Earth*; cf. pp. 319 sq. According to J. L. E. Dreyer, *A history of astronomy from Thales to Kepler,* 2nd ed., New York, 1953, p. 348, Gilbert, in his posthumous work, *De mundo nostro sublunari philosophia nova* (Amstelodami, 1651), " appears to hesitate between the system of Tycho and Copernicus." This is not quite exact, since Gilbert, in contradistinction to Tycho Brahe, (a) asserts the rotation of the earth which Tycho Brahe rejects, and (b) denies the existence of a sphere of fixed stars, and even the finitude of the universe still taught by Brahe. Thus Gilbert tells us that though the majority of the philosophers placed the earth in the center of the world, there is no reason to do so (l. 2, cap. II, *De telluris loco,* p. 115): " Telluris

vero globum in centro universi manentem omnis fere philosophorum turba collocavit. At si motum aliquem habuerit praeter diurnam revolutionem (ut nonnulli existimant) erronem etiam illam oportet esse; sin in suo sede volveretur tantum, non in circulo, planetarum ritu moveretur. Non tamen inde, aut ullis aliunde depromptis rationibus, certo persuadetur eam in universae rerum naturae centro, aut circa centrum, permanere." He adds, indeed (*ibid.*, p. 117), that "Non est autem quo persuaderi possit in centro universi magis terram reponi quam Lunam, quam Solem; nec ut in motivo mundo horum unum in centro sit, necesse esse," and that, moreover, the world itself has no center (p. 119).

On the other hand, though he puts the sun and not the earth in the center of the moving world (p. 120): "locus telluris non in medio quia planetae in motu circulari tellurem non observant, tanquam centrum motionum, sed Solem magis," and tells us that the sun (p. 158) "maximam vim egendi et impellendi habet, qui etiam motivi mundi centrum est," he does not tell us outright that the earth belongs to this "moving world" of the planets.

Though he quotes Copernicus and even tells us that Copernicus erred in ascribing to the earth three motions, instead of two (around its axis and around the sun), the third one, that which, according to Copernicus, turned the axis of the earth in order to keep it pointing always in the same direction being not a motion at all, but lack of it (p. 165): "Tertius motus a Copernico inductus non est motus omnino, sed telluris est directio stabilis," he does not assert the truth of the heliocentric world-view.

He tells us, indeed (l. I, cap. XX, *De vacuo separato*), that the Aristotelian objections against the void are worthless, that things can just as well move in the void space as remain immobile in it and that the earth can very well be a planet and turn around the sun like the others; that, nevertheless, he does not want to discuss this question (l. I, cap. XX, *De vacuo separato*, p. 49): "Cujus rei veritatem sic habeto. Omnia quiescunt in vacuo posita; ita quies plurimis globis mundi. At nonnulli globi et infinitis viribus et actu aliorum corporum aguntur circa quaedam corpora, ut planetae circa Solem, Luna circa Tellurem et erga Solem.

"Quod si Sol in medio quiescit ut Canis, ut Orion, ut Arcturus, tum planetae, tum etiam tellus, a Sole aguntur in orbem, consentientibus propter bonum ipsis globorum formis: si vero tellus in medio quiescat (de cujus motu annuo non est huius loci disceptare) aguntur circa ipsam cetera moventia."

It is possible, of course, that Gilbert really considered that the discussion of the annual motion of the earth was out of place in a book devoted to the development of a new philosophy of our sublunar world. Yet it is difficult to admit that, if he was fully convinced of the truth of the Copernican astronomy, he would so consistently avoid saying it, even when asserting its daily rotation, as, for instance in chap. VI of book II of the *Philosophia nova* (p. 135): "Terram circumvolvi diurno motu, verisimile videtur: an vero circulari aliquo motu annuo cietur, non hujus est loci inquirere." It seems, thus, that Gilbert was either not very much interested in the problem, or sceptical about the possibility to reach a solution and that he hesitated between an improved Copernicanism (such as Kepler's) and an improved Tycho Brah-ism (such as Longomontanus').

CHAPTER III

1. In pointing out the analogy between Kepler's views and those of some modern scientists and philosophers of science I am *not* committing an anachronism: epistemology and logic are, indeed, nearly as old as science itself and empiricism or positivism are by no means new inventions.
2. The sun represents, symbolizes, and perhaps even embodies God the Father, the stellar vault, the Son, and the space in between, the Holy Ghost.
3. Cf. *De stella nova in pede Serpentarii,* cap. XXI, pp. 687 (*Opera omnia,* ed. Frisch, vol. II, Frankofurti et Erlangae, 1859). The *De stella nova* was published in 1606.
4. *Ibid.,* p. 688.
5. *Ibidem.*
6. *Ibidem.*
7. *Ibidem.*
8. *Ibidem.*
9. A perfectly reasonable assumption, and quite analogous to that of contemporary astronomy, about the distribution of galaxies.
10. *De stella nova,* p. 689.
11. *Ibidem.*
12. *Ibidem.*
13. The sky being "above" us, the stars are "elevated" in respect to us; thus to place them at a greater distance from us (or the centre of the world) is to give them a greater "elevation."
14. *Ibid.,* pp. 689 sq.

15. The absence of stellar parallaxes imposes a *minimum* to the distance separating us from the fixed stars.
15a. Marcus Manilius, a Stoic, who lived in the Augustan age, author of a great astrological poem, *Astronomicon libri quinque*, which was edited by Regiomontanus in Nürnberg in 1473.
16. *Ibid.*, p. 690.
17. *Ibidem.*
18. Two minutes is the magnitude of the *visible* diameter of a star for the unassisted eye.
19. *Ibidem.*
20. *Ibid.*, p. 691.
21. *Ibidem.*
22. *Ibidem.*
23. *Ibidem.*
24. J. Kepler, *Dissertatio cum Nuntio Sidereo nuper ad mortales misso a Galileo Galilei*, p. 490 (*Opera omnia*, vol. II), Frankoforti et Erlangae, 1859. Wacherus = the Imperial Councillor Wackher von Wackenfels who was the first to inform Kepler about the discoveries of Galileo. Brutus = the Englishman Edward Bruce who was a partisan of Giordano Bruno and who, some years before (Nov. 5, 1603), sent to Kepler a letter (from Venice) in which he expressed his belief in the infinity of the world; according to Bruce fixed stars were suns surrounded by planets like our sun, and, like our sun, endowed with a rotational motion. Bruce's letter is quoted by Frisch, *Opera omnia*, vol. II, p. 568, and published by Max Caspar in his edition of Kepler (Johannes Kepler, *Gesammelte Werke*, vol. IV, p. 450, München, 1938).
25. The fixed stars, as seen by a Galilean telescope, do not appear as light-points; they still have visible dimensions; cf. *supra*, p. 191.
26. *Epitome astronomiae Copernicanae*, liber I, pars II, p. 136 (*Opera omnia*, vol. VI, Frankoforti et Erlangae, 1866).
27. *Ibidem.*
28. *Ibidem.*
29. *Ibid.*, p. 137.
30. *Ibidem.*
31. *Ibid.*, p. 138.
32. *Ibidem.*
33. *Ibidem.*
34. *Ibidem.*
35. *Ibidem.*
36. *Ibid.*, p. 139.
37. Contemporary cosmology, on the other hand, seems to have recognized

the value of the old doubts about the possibility of an actually infinite world, and turned back to a finitist conception.

38. That is the conception ascribed by Plutarch (or Pseudo-Plutarch) to the Stoics.
39. *Ibid.*, p. 139.
40. Cf. my paper, "Le vide et l'espace infini au XIVème siècle," *Archives d'histoire doctrinale et littéraire du Moyen-Age,* XVII, 1949.

CHAPTER IV

1. Galileo Galilei, *Sidereus nuncius* . . . Venetiis, 1610; there is an English translation by E. S. Carlos, *The sidereal messenger,* London, 1880. Large parts of this translation are reprinted in Harlow Shapley and Helen E. Howarth, *A source book in astronomy,* New York, 1929. Though not using this translation I refer to it whenever possible. The expresssion *Sidereus Nuncius* was used by Galileo as meaning: the *message* of the stars. Yet Kepler understood it as meaning: the messenger of stars. This mistranslation became generally accepted and was corrected only in the recent edition of the *Nuncius* by Mrs. M. Timpanaro-Cardini, Florence, 1948.
2. Cf. *Sidereus nuncius,* pp. 59 sq. (*Opere, Edizione Nazionale,* v. III, Firenze, 1892), *Source book,* p. 41.
3. On the discovery of the telescope cf. Vasco Ronchi, *Galileo e il cannochiale,* Udine, 1942, and *Storia della luce,* 2 ed., Bologna, 1952.
4. *Sidereus nuncius,* p. 75, *Source book,* p. 46.
5. *Ibid.*, p. 76.
6. *Ibid.*, p. 78.
7. Galileo Galilei, *Letter to Ingoli,* p. 526. *Opere, Ed. Naz.*, vol. VI, Firenze, 1896.
8. It is interesting to note that the conception according to which heavenly bodies are inhabited is referred to by Galileo as "commonly held."
9. *Letter to Ingoli,* p. 525.
10. *Ibid.*, p. 518.
11. Galileo Galilei, *Dialogo sopra i due massimi sistemi del mondo* (*Opere, Ed. Naz.*, vol. VII), p. 44; Firenze, 1897; cf. also p. 333. The *Dialogue* is easily available now in the excellent modernization of the old Salusbury translation by Professor Giorgio di Santillana, Galileo Galilei, *Dialogue on the great world systems,* Chicago, 1953, as well as in the new translation by Stillman Drake, Galileo Galilei, *Dialogue concerning the two*

chief world systems, Ptolemaic and Copernican, Berkeley and Los Angeles, 1953.

12. *Dialogo,* p. 306.
13. *Letter to Ingoli* (*Opere,* vol. VI), pp. 518, 529.
14. *Dialogo, loc. cit.*
15. Cf. *Letter to Liceti,* of February 10, 1640; *Opere,* vol. XVIII, pp. 293 sq., Firenze, 1906.
16. Cf. R. Descartes, *Principia philosophiae,* part II, §4, p. 42. (*Oeuvres,* ed. Adam Tannery, vol. VIII, Paris, 1905.)
17. *Principia philosophiae,* pt. II, §10, p. 45.
18. *Ibid.,* §11, p. 46.
19. *Ibid.,* §13, p. 47.
20. *Ibid.,* §13, p. 47.
21. *Ibid.,* §16, p. 49.
22. *Ibid.,* §21, p. 52.
23. *Ibid.,* §22, p. 52.
24. *Ibidem.*
25. *Principia philosophiae,* p. I, §26, p. 54.
26. *Ibid.,* §27, p. 55.
27. *Ibidem.*
28. *Principia philosophiae,* p. III, §1, p. 80.
29. *Ibid.,* §2, pp. 81 sq.

CHAPTER V

1. Cf. Miss Marjorie H. Nicolson, "The early stages of Cartesianism in England," *Studies in Philology,* vol. XXVIII, 1929. Henry More accepted Cartesian physics, though only partially, and the Cartesian rejection of substantial forms, but he never abandoned his belief in the existence, and action, of "spiritual" agents in nature and never adopted the Cartesian strict opposition of matter — reduced to extension — to spirit, defined by self-consciousness and freedom. Henry More, accordingly, believes in animals 'having souls and in souls' having a non-material extension; cf. also Miss Nicolson's *The breaking of the circle,* Evanston, Ill., 1950.
2. These letters were published by Clersellier in his edition of the correspondence of Descartes (*Lettres de M. Descartes où sont traittées les plus belles questions de la morale, de la physique, de la médecine et des mathématiques* . . . Paris, 1657) and republished by Henry More

himself (with a rather angry preface) in his *Collection of severall philosophical writings* of 1662. I am quoting them according to the text of the Adam-Tannery edition of the works of Descartes (*Oeuvres*, vol. v, Paris, 1903).

3. *Letter to Descartes,* II-XII, 1648, pp. 238 sq.
4. In this work, written in 1646, he shows himself an enthusiastic follower of the Lucretian-Brunonian doctrine of the infinity of worlds; cf. Lovejoy, *op. cit.*, pp. 125, 347.
5. On Gassendi see K. Lasswitz, *op. cit.*, and R. P. Gaston Sortais, *La philosophie moderne, depuis Bacon jusqu'à Leibniz,* vol. II, Paris, 1922; also *Pierre Gassendi, sa vie et son oeuvre,* Paris, 1955. Gassendi is not an original thinker and does not play any role in the discussion I am studying. He is a rather timorous mind and accepts, obviously for theological reasons, the finitude of the world immersed in void space; yet, by his revival of Epicurean atomism and his insistence upon the existence of the void, he undermined the very basis of the discussion, that is, the traditional ontology which still dominated the thought not only of Descartes and More but also of Newton and Leibniz.
6. *Letter to Descartes,* p. 242.
7. In the Cartesian world vortices which surround fixed stars limit each other and prevent each other from spreading and dissolving under the influence of centrifugal force; if they were limited in number, and therefore in extension, then, first the outermost ones and then all the others would be dispersed and dissipated.
8. *Letter to Descartes,* p. 242.
9. Namely, by arguments based upon the consideration of God's omnipotence.
10. *Descartes to Henry More,* 5, II, 1649, pp. 267 sq.
11. *Ibid.*, pp. 269 sq.
12. *Ibid.*, p. 274.
13. *Ibid.*, p. 275.
14. *Second letter of H. More to Descartes,* 5, III, 49; pp. 298 sq.
15. *Ibid.*, pp. 304 sq.
16. *Ibid.*, p. 305.
17. *Ibid.*, p. 302. More's argument against Descartes is a re-edition of Plotinus' argument against Aristotle.
18. *Ibid.*, p. 312; cf. *supra.*
19. *Second letter of Descartes to Henry More,* 15, IV, 1649; pp. 340 sq.
20. *Ibid.*, p. 342.
21. *Ibid.*, p. 343.

22. Such was, in any case, the opinion of Pascal. Yet, after all, what is the God of a philosopher supposed to be if not a philosophical God?
23. *Ibid.*, p. 344.
24. *Ibid.*, p. 345.
25. Dated the 23rd of July, 1649 (*Oeuvres*, vol. v, pp. 376 sq.).
26. At least, he started writing an answer — in August 1649 — though he did not send it to Henry More.
27. Dated the 21st of October, 1649, vol. v, pp. 434 sq.
28. It is possible, of course, that, as he went to Sweden on Sept. 1, 1649 and died there on Feb. 11, 1650, Descartes did not receive this last letter of Henry More.
29. Cf. my *Essai sur les preuves de l'existence de Dieu chez Descartes*, Paris, 1923, and "Descartes after three hundred years," *The University of Buffalo Studies*, vol. XIX, 1951.

CHAPTER VI

1. Henry More has not received the monographical treatment to which he is undoubtedly entitled. On him, and on the Cambridge Platonists in general, cf. John Tulloch, *Rational theology and Christian philosophy in England in the XVIIth century*, vol. II, Edinburgh and London, 1874; F. J. Powicke, *The Cambridge Platonists*, London, 1926; J. H. Muirhead, *The Platonic tradition in Anglo-Saxon philosophy*, London, 1931; T. Cassirer, *Die Platonische Renaissance in England und die Schule von Cambridge*, Leipzig, 1932; English translation: *The Platonic Renaissance in England and the Cambridge School*, New Haven, 1953. Selections of the philosophical writings of Henry More (namely from *The antidote against atheism*, *The immortality of the soul*, and the *Enchiridium metaphysicum* in translation) were published in 1925 by Miss Flora J. Mackinnon with an interesting introduction, valuable notes, and an excellent bibliography: *Philosophical writings of Henry More*, New York, 1925. Cf. Marjorie Nicolson, *Conway letters, the correspondence of Anna, Viscountess Conway, Henry More and their friends, 1642-1684*, London, 1930; Markus Fierz, "Ueber den Ursprung und Bedeutung der Lehre Newtons vom absolutem Raum," *Gesnerus*, vol. XI, fasc. 3/4, 1954; Max Jammer, *Concept of space*, Harvard Univ. Press, Cambridge, Mass., 1954. Both Markus Fierz and Max Jammer seem to me to exaggerate the real influence of cabalist space conceptions on Henry More (and his predecessors). In my opinion, it was a typical

case of reprojection into the past of modern conceptions in order to back them up by sacred or venerable authorities; yet, as we know, misunderstanding and misinterpretation play an important part in the history of thought. It seems to me, moreover, that Fierz and Jammer themselves are not quite innocent of the sin of retroprojection, forgetting that space conceptions formed before the invention of geometry were not, and could not, be identical or even similar to the conceptions devised after this momentous event.

2. Henry More, *An antidote against atheisme, or an appeal to the natural faculties of the minde of man, whether there be not a God,* London, 1652; second ed. corrected and enlarged, London, 1655; third edition, corrected, and enlarged, "with an Appendix thereunto annexed," London, 1662. I am quoting this edition as given in Henry More's *Collection of severall philosophical writings,* London, 1662.
3. Henry More, *The immortality of the soul, so farre forth as it is demonstrable from the knowledge of nature and the light of reason,* London, 1669; second edition in the *Collection of severall philosophical writings* of 1662. It is this edition that I am quoting.
4. Henricus Morus, *Enchiridium metaphysicum sive de rebus incorporeis succincta et luculenta dissertatio,* Londini, 1671.
5. Henry More, *An antidote against atheism,* book I, cap. IV, §3, p. 15.
6. Henry More, *The immortality of the soul,* b. I, c. II, axiom IX, p. 19.
7. Cf. R. Zimmerman, "Henry More und die vierte Dimension des Raumes," *Kaiserliche Akademie der Wissenschaften,* Philosophisch-historische Klasse, Sitzungsberichte, Bd. 98, pp. 403-sq., Wien, 1881.
8. Henry More, *The immortality of the soul,* b. I, c. II, §11, p. 20.
9. *Ibid.,* 6, I, c. III, §§1 and 2, pp. 21 sq.
10. Axiom IX (b. I, c. II, p. 19) tells us that "There are some Properties, Powers and Operations, immediately appertaining to a thing, of which no reasons can be given, nor ought to be demanded, nor the Way or Manner of the cohesion of the Attribute with the subject can by any means be fancied or imagined."
11. Cf. William Gilbert, *De magnete,* ch. XII, p. 308: "The magnetic force is animate, or imitates the soul; in many respects it surpasses the human soul while that is united to an organic body."

11a. Cf. also Markus Fierz, *op. cit.,* pp. 91 sq.

12. Henry More, *The immortality of the soul,* b. III, c. XII, §1, p. 193.
13. *Ibid.,* preface, §12, p. 12.
14. *An antidote against atheism,* c. II, c. II, §1, p. 43.
15. *Ibid., Appendix* (of 1655), cap. VII, §1, p. 163.
16. *Ibidem.*

17. *Ibid.*, §§4, 5, 6, pp. 164 sq.
18. *Enchiridium metaphysicum,* part I, cap. VI, v. 42.
19. *Ibidem.*
20. *Ibidem.*
21. *Ibid.*, cap. VI, 4, p. 44.
22. *Ibid.*, cap. VI, 11, p. 51.
23. *Ibid.*, cap. VII, 3, p. 53.
24. This definition is given by Descartes in the *Principia philosophiae,* part II, §25.
25. *Enchiridium metaphysicum,* cap. VII, 7, p. 56.
26. *Ibid.*, c. VII, 6, p. 55.
27. *Ibidem.*
28. Ibidem.
29. *Ibidem.*
30. *Ibid.*, c. VIII, 6, p. 68.
31. *Ibid.*, c. VIII, 7, p. 69.
32. *Ibid.*, c. VIII, 8, pp. 69 sq.
33. *Ibid.*, c. VIII, 9, p. 70.
34. *Ibid.*, c. VIII, 10, p. 71.
35. *Ibid.*, c. VIII, 11, p. 72.
36. *Ibid.*, c. VIII, 12, p. 72.
37. *Ibidem.*

CHAPTER VII

1. Cf. Nicolas Malebranche, *Méditations chrétiennes,* méd. IX, §9, p. 172, Paris, 1926. On Malebranche cf. H. Gouhier, *La philosophie de Malebranche,* Paris, 1925.
2. *Ibidem.*
3. *Ibid.*, §10, p. 173.
4. *Ibid.*, §8, pp. 171 sq.
5. *Ibid.*, §11, p. 174.
6. *Ibid.*, §12, pp. 174 sq.
7. Cf. Malebranche, *Correspondance avec J. J. Dortous de Mairan,* ed. nouvelle, précédée d'une introduction par Joseph Moreau, Paris, 1947.
8. Cf., e. g., the already quoted book of E. A. Burtt, *The metaphysical foundations of modern physical science,* New York, 1925; second ed., London, 1932.
9. Cf. *Sir Isaac Newton's mathematical principles of natural philosophy,*

translated into English by Andrew Motte in 1729, the translation revised by Florian Cajori, p. 6, Berkeley, Calif., 1946.

10. *Ibid.*, p. 8.
11. *Ibidem.*
12. *Ibid.*, p. 6.
13. *Ibidem.*
14. *Ibidem.*
15. *Ibidem.*
16. *Ibidem.*
17. *Ibidem.*
18. *Ibid.*, p. 7. The example of the sailor is discussed by Descartes in the *Principia philosophiae,* II, 13, 32.
19. *Ibid.*, p. 8.
20. His pupil, Dr. Clarke, will indeed do it; cf. *infra,* p. 275.
21. *Ibid.*, p. 9.
22. *Ibid.*, p. 10.
23. *Ibidem.*
24. *Ibid.*, p. 11. As against Descartes, *Principia,* II, 13.

24a. Cf. Ernst Mach, *The science of mechanics,* Chicago, 1902, pp. 232 sq.; cf. also Max Jammer, *op. cit.*, pp. 104 sq.; 121 sq.; 140 sq.

25. *Ibid.*, p. 12.
26. *Ibid.*, book III, *The system of the world,* Lemma IV, cor. III, p. 497.
27. *Ibid.*, book III, *The system of the world,* prop. V, theorem VI, scholium, cor. III, p. 414.
28. *Ibid.*, cor. IV, p. 415.
29. As a matter of fact, they have been listed also by Boyle and Gassendi who, in contradistinction to Descartes, insist on impenetrability as an irreducible property of body distinct from mere extension.
30. *Ibid.*, rule III, pp. 398 sq. The text I am referring to appeared in the *second* edition of the *Principia*; yet, as it represents the fundamental views of Newton which inspired his whole system, I feel it necessary to quote it here. On the difference between the *first* and the subsequent editions of the *Principia,* cf. my papers "Pour une édition critique des oeuvres de Newton," *Revue d'Histoire des Sciences,* 1955, and "Expérience et hypothèse chez Newton," *Bulletin de la Société Française de Philosophie,* 1956.
31. *Ibid.*, book I, sect. XI, prop. LXIX, schol., p. 192.
32. Cf. my *Études Galiléennes.* II, *La loi de la chute des corps,* and III, *Galilée et la loi d'inertie.*
33. *Ibid.*, *loc. cit.*
34. *Four Letters from Sir Isaac Newton to the Reverend Dr. Bentley,*

Letter II (Jan. 17, 1692-93), p. 210, London, 1756; reprinted in *Opera omnia*, ed. by Samuel Horsley, 5 vols., London, 1779-85 (vol. IV, pp. 429-442), and also in the *Works* of R. Bentley, vol. III, London, 1838. I am quoting this edition.

35. Letter III (Feb. 25, 1692-93), *ibid.*, p. 211.
36. *Eight sermons preach'd at the Honourable Robert Boyle lecture in the first year MDCXCII*, By Richard Bentley, Master of Arts, London, 1693. The first sermon proves *The folly of atheism and . . . Deism even with respect to the present life*, the second demonstrates that *matter and motion cannot think*, the third, fourth and fifth present *A confutation of atheism from the structure of the human body*, the sixth, seventh and eighth, forming the second part of the work, *A con futation of atheism from the origin and frame of the world.* I am quoting the last edition (*Works*, v. III) of this book that has seen nine of them in English, and one in Latin (Berolini, 1696); cf. Part II, sermon VII (preached Nov. 7th, 1692), pp. 152 sq.
37. *Ibid.*, p. 154.
38. *Ibid.*, p. 157.
39. *Ibid.*, pp. 162 sq.
40. *Ibid.*, p. 163.
41. *Letters from Sir Isaac Newton to the Reverend Dr. Bentley*, Letter I, pp. 203 sq.
42. *A confutation of atheism from the origin and frame of the world*, p. 165.
43. *Ibid.*, p. 170.
44. *Ibid.*, pp. 175 sq.
45. On the cosmical optimism of the XVIIIth century, cf. Lovejoy, *op. cit.*, pp. 133 sq.; E. Cassirer, *Die Philosophie der Aufklärung*, Tübingen, 1932.

CHAPTER VIII

1. Joseph Raphson is chiefly known as the author of the violently pro-Newtonian *Historia Fluxionum, sive Tractatus Originem et Progressum Peregregiae Istius Methodis Brevissimo Compendio (Et quasi synoptice) Exhibens*, Londini, 1715.
2. *Analysis Æquationum Universalis seu ad Aequationes Algebraicas Resolvendas Methodus Generalis et Expedita, Ex nova Infinitarum Serierum Methodo, Deducta et Demonstrata.* Editio *secunda* cui accedit *Appendix* de *Infinito Infinitarum Serierum* progressu ad *Equationum Algebraicarum Radices* eliciendas. Cui etiam Annexum est De SPATIO REALI seu ENTE INFINITO conamen Mathematico Metaphysicum, Authore

Josepho Raphson *A.M.* et Reg. Soc. Socio., Londini, 1702. The first edition of J. Raphson's work, without the above-mentioned appendices, appeared in 1697.

3. *De ente infinito,* cap. iv, p. 67.
4. Cf. *infra,* pp. 193, 196.
5. *De ente infinito,* cap. iv, pp. 57 sq.
6. *Ibid.,* pp. 70 sq.
7. *Ibid.,* cap. v, p. 72.
8. *Ibid.,* Def. I.
9. *Ibid.,* Scholium, p. 73.
10. *Ibidem.*
11. *Ibid.,* pp. 74 sq.
12. *Ibid.,* Scholium, p. 76. On the space theories of the *Cabala* cf. Max Jammer, *op. cit.,* pp. 30 sq.
13. *Ibid.,* corollarium.
14. *Ibidem.*
15. *Ibid.,* p. 78.
16. *Ibid.,* p. 80.
17. *Ibid.,* cap. vi, p. 82.
18. *Ibid.,* p. 83.
19. *Ibid.,* pp. 83 sq.
20. *Ibid.,* p. 85.
21. *Ibid.,* pp. 90 sq.
22. *Ibid.,* p. 91.
23. *Ibid.,* p. 91.
24. *Ibid.,* pp. 91 sq.
25. *Ibid.,* p. 92.
26. *Ibid.,* p. 93.
27. *Ibid.,* p. 95.

CHAPTER IX

1. Strange as it may seem, the adjunction of these "queries," numbered 17 to 23, to the Latin edition of the *Opticks* in 1706 seems to have escaped the attention of Newton's historians who, usually, attribute these queries to the second (English) edition of 1617 of his *Opticks.* Thus, for instance, L. T. More, *Isaak Newton,* New York-London, 1934, p. 506, note: "A second edition (octavo) bears the advertisement 1717. It was published in 1718. . . . The number of new Queries added

begins with the seventeenth." Leon Bloch's *La philosophie de Newton*, Paris, 1908, is an honorable exception to the afore-mentioned rule; and today, Mr. H. G. Alexander, editor of *The Leibniz-Clarke correspondence*, Manchester University Press, 1956.

2. *Philosophical principles of natural religion* by George Cheyne, M. D. and F. R. S., London, 1705. The second edition of Cheyne's book, published under the title *Philosophical principles of religion, natural and revealed*, London, 1615, "corrected and enlarged," contains two parts: Part I, "containing the *Elements* of *Natural Philosophy* and the Proofs of NATURAL RELIGION arising from them," and a Part II, "containing the *Nature* and *Kinds of Infinities*, the *Arithmetick* and *Uses*, and the *Philosophick Principles* of *Reveal'd Religion*, now first published." Strangely enough the common title page, as well as that of the second part, bears the date 1715, whereas that of the first part, the date 1716. As a matter of fact, or at least according to David Gregory who held this information from Newton himself, it was the publication by Dr. Cheyne of his *Fluxionum methodus inversa sive quantitatum fluentium leges generales*, London, 1703 (rather sharply criticized by A. De Moivre in his *Animadversiones in Dr. G. Cheyne's Fluxionum methodus* . . . London, 1704), which prompted Newton to publish the *Two treatises on the species and magnitudes of curvilinear figures*, that is, *The quadrature of curves* and *The enumeration of the lines of the third order*; (cf. *David Gregory, Isaak Newton and their circle*, Extracts from David Gregory's *Memoranda*, edited by W. G. Hiscock, pp. 22 sq., Oxford, 1937). In the selfsame *Memoranda* under the date of December 21, 1705, we find also the following, very interesting passage (*ibid.*, pp. 29-30): "Sir Isaak Newton was with me and told me that he had put 7 pages of Addenda to his Book of Lights and Colours in this new latin edition of it. He has by way of quaere explained the explosion of Gun powder, all the chief Operations of Chymistry. He has shewed that Light is neither a communication of motion nor of a Pressure. He inclines to believe it to be projected minute bodys. He has explained in those Quaerys the double Refraction in Iseland Crystall. His Doubt was whether he should put the last Quaere thus. *What the space that is empty of bodies is filled with*. The plain truth is that he believes God to be omnipresent in the literal sense. And that as we are sensible of Objects where their images are brought within the brain, so God must be sensible of every thing being intimately present with every thing: for he supposes that as God is present in space where there is no body, he is present in space where a body is also present. But if this way of proposing this his notion be too bold, he thinks of doing it thus. *What cause did the Ancients assign of*

Gravity. He believes that they reckoned God the Cause of it, nothing else, that is no body being the cause; since every body is heavy.

"Sir Isaak believes that the Rays of Light enter into the composition of most Natural Bodies that is the small particles that are projected from a lucid body in form of Rays. As plain this may be the case with most combustible, inflammable bodies." On the relations of light and matter according to Newton cf. Helène Metzger, *Newton, Stahl, Boerhaave et la doctrine chimique,* Paris, 1930.

3. *Optice* . . . l. III, qu. 20, pp. 312 sq.; London, 1706; qu. 28 of the English edition; cf. I. Bernard Cohen's edition of the *Opticks,* New York, 1952, p. 369. As the English edition certainly gives the original text of Newton himself, I will quote this latter giving first the page numbers of the Latin, and then those of the afore-mentioned edition.
4. *Ibid.*, pp. 322 sq.; pp. 375-76. The existence of various "impellent" and "repellent" forces acting between the "particles" of bodies is already asserted by Newton in the preface of the *Principia.*
5. *Ibid.*, p. 376.
6. *Ibid.*, p. 335; pp. 388 sq.
7. *Ibid.*, p. 335 sq.; pp. 389 sq.
8. *Ibid.*, p. 337; p. 394.
9. *Ibid.*, pp. 337 sq.; pp. 394 sq.
10. *Ibid.*, pp. 338 sq.; pp. 395-396.
11. *Ibid.*, pp. 340 sq.; pp. 397 sq.

11a. The reasoning is, of course, utterly false and it is rather astonishing that Newton could have made it and that neither he himself nor his editors noticed this falsehood.

12. *Ibid.*, p. 343; p. 399.
13. *Ibid.*, pp. 343 sq.; p. 400.
14. *Ibid.*, p. 345; p. 402.
15. *Ibid.*, p. 346; p. 403.

CHAPTER X

1. George Berkeley, *Principles of human knowledge,* §110; p. 89 (*The works of George Berkeley Bishop of Cloyne,* ed. by A. A. Luce and T. E. Jessop, vol. I, Edinburgh, 1949).
2. *Ibid.*, §111, p. 90.
3. *Ibid.*, §117, p. 94.
4. On the 18th of February 1673, Roger Cotes wrote to Newton (cf.

Correspondence of Sir Isaak Newton and Professor Cotes . . . ed. J. Edleston, London, 1850, pp. 153 sq.): ". . . I think it will be proper [to] add something by which your book may be cleared from some prejudices which have been industriously laid against it. As that it deserts mechanical causes, is built upon miracles and recurrs to Occult qualities. That you may not think it unnecessary to answer such Objections you may be pleased to consult a Weekly Paper called *Memoires of Literature* and sold by Ann Baldwin in Warwick-Lane. In the 18th Number of ye second Volume of those Papers which was published May 5th, 1712, you will find a very extraordinary letter of Mr. Leibnitz to Mr. Hartsoeker which will confirm what I have said." Indeed, in this letter, dated Hanover, February 10, 1711, Leibniz who, as a matter of fact already had attacked Newton in his *Théodicée* (*Essai de Théodicée, Discours de la Conformité de la Foi avec la Raison*, §19, Amsterdam, 1710) assimilated the Newtonian gravity to an "occult quality," so "occult" that it could never be cleared up even by God. It is well known that neither Leibniz nor Huygens had ever accepted the Newtonian conception of gravitation, or attraction. Cf. René Dugas, *Histoire de la mécanique au XVIIe siècle*, Neuchatel, 1954, cap. XII, *Retour au Continent*, pp. 446 sq. and cap. XVI, *Réaction des Newtoniens*, pp. 556 sq.

4a. In the first line, Henry More and Joseph Raphson.

5. Cf. *Mathematical principles of natural philosophy*, translated into English by Andrew Motte in 1729. The translation revised . . . by Florian Cajori, General Scholium, pp. 543 sq., Berkeley, Cal., 1946.

6. *Ibid.*, pp. 544 sq.

7. *Ibid.*, p. 545.

8. *Ibidem.*

9. *Ibidem.*

10. *Ibid.*, p. 546.

11. *Ibid.*, pp. 546 sq.

12. Professor Cajori follows Andrew Motte in translating Newton's *fingo* by *frame*. It seems to be that the old term *feign* (used by Newton himself) is both more correct and more expressive.

13. *Principles*, preface, p. xx.

14. *Ibid.*, p. xxix.

15. *Ibid.*, p. xxvii.

16. *Ibid.*, pp. xxxi sq.

17. *Principles*, p. 547. On the XVIIth century conception of "spirit" cf. E. A. Burtt, *op. cit.*, and A. J. Snow, *Matter and gravity in Newton's philosophy*, Oxford, 1926.

NOTES

CHAPTER XI

1. Wilhelmine Caroline, later Queen Caroline, was born Princess of Brandenburg-Anspach and in 1705 became the wife of George Augustus, Electoral Prince of Hanover. It was as Princess of Hanover that she became intimate with Leibniz; as Leibniz put it himself, she "inherited" him from Sophie Charlotte of Prussia.
2. Cf. "An extract of a letter written in November 1715," §§3 and 4, published in *A Collection of papers, which passed between the late learned Mr. Leibnitz and Dr. Clarke. In the years 1715 and 1716 Relating to the Principles of Natural Philosophy and Religion. With an Appendix*, pp. 3 and 5, London, 1717. Leibniz writes, of course, in French, and Clarke, in English. But he accompanies the publication of the originals by a translation of Leibniz's "papers" into English (probably made by himself) and of his own "replies" into French (probably made by the Abbé Conti). Moreover, he adds to the text a series of footnotes with references to relevant passages in Newton's writings. This polemic is now available in the excellent edition of G. H. Alexander, *The Leibniz-Clarke correspondence,* Manchester Univ. Press, 1956; cf. also René Dugas, *La mécanique au XVII siècle,* cap. XVI, §3, pp. 561 sq.
3. The choice of Dr. Samuel Clarke was rather obvious. Dr. Clarke, Rector of St. James', Westminster, was not only a philosophical theologian — in 1704-5 he gave the Boyle Lectures — but also was former chaplain of Queen Anne, removed, to say the truth, from this charge for lack of orthodoxy (he was practically an Arian). However, after Queen Anne's death he became an intimate of Princess Caroline with whom, at her request, he had weekly philosophical conversations in which other gentlemen interested in discussing philosophical problems participated. Thus it was only natural that, as Des Maizeaux tells us in the preface to his own French re-edition of the *Collection of papers* (*Recueil de diverses pièces sur la philosophie, la religion naturelle, l'histoire, les mathématiques etc.*, 2 vols., Amsterdam, 1720, p. II): "Madame la Princesse de Galles, accoutumée aux Recherches Philosophiques les plus abstraites et les plus sublimes fit voir cette Lettre à M. Clarke et souhaita qu'il y répondit. . . . Elle envoyait à M. Leibniz les Réponses de M. Clarke et communiquait à M. Clarke les nouvelles difficultés, ou les Instances de M. Leibniz." Indeed, Dr. Clarke as an intimate friend of Sir Isaac, and a Newtonian of long standing, could be relied upon to represent the philosophical views of his master.

In my opinion we must go even farther: it is utterly unconceivable that Clarke should accept the role of philosophical spokesman (and defender) of Newton without being entrusted by the latter to do it, nay, without having secured the collaboration of the great man, at least in the form of approval.

I am, thus, morally certain that Clarke communicated to Newton *both* Leibniz's letters and his own replies to them. It is indeed unthinkable that in the midst of his bitter fight with Leibniz about the priority of the invention of the calculus, Newton who "aided" both Keill and Raphson in their attacks against Leibniz, as he "aided" Des Maiseaux some years later in the preparation of his edition of the "*Collection of papers*" (the second volume of his edition carries the history of the calculus controversy by publishing translations of selected pieces of the *Commercium epistolicum*), should remain aloof and disinterested in the face of an assault upon his religious view and an accusation, practically, of atheism, by the selfsame Leibniz. As a matter of fact, the Princess of Wales informed Leibniz (Caroline to Leibniz, Jan. 10, 1716, in O. Klopp, *Die Werke von Leibniz*, Hanover, 1864-84, vol. XI, p. 71, quoted in *The Leibniz-Clarke correspondence*, Manchester Univ. Press, 1956, p. 193) that he was right in his supposition that these letters were not written without the advice of Newton. Strange as it may seem, the importance of Clarke's papers as representing *literally* the metaphysical views of Newton has never been recognized, with the result that their study was completely neglected by the historians both of Newton and of Leibniz. Thus, for instance, L. T. More, *op. cit.*, p. 649: "It seems probable that Newton was even more exasperated by Leibniz's attack on the anti-Christian influence of the *Principia* than by the controversy over the invention of the calculus. To justify himself he guided Des Maizeaux in preparing for publication the long debate between Leibniz and Samuel Clarke on the religious significance of the Newtonian Philosophy. For this purpose he gave to the author the documents relating to the controversy, and assisted him in preparing an historical preface which reviewed the whole affair."

4. Cf. *supra*, pp. 181-89.

4a. As a matter of fact (cf. *supra*, p. 209) Newton, at least once, identified space with God's *sensorium*.

5. "Dr. Clarke's first reply," *A collection of papers . . .*, pp. 15 sq.

5a. The Socinians did not believe in predestination, nor in the Trinity.

6. "Mr. Leibniz's second paper," *ibid.*, p. 25.

7. *Ibid.*, p. 33.

8. Especially his allusion to Socinianism, because, as a matter of fact both

Sir Isaac Newton and Dr. Samuel Clarke were much nearer to Socinianism than to the teaching of the Established Church: neither of them, indeed, accepted the Trinitarian conception of God; they were both — as also John Locke — Unitarians; cf. H. McLachlan, *The religious opinions of Milton, Locke and Newton,* Manchester, 1941. On Newton's metaphysical and religious views, cf. Helène Metzger, *Attraction universelle et religion naturelle,* Paris, 1938, and E. W. Strong, "Newton and God," *Journal of the History of Ideas,* vol. XIII, 1952.

9. Or, at least, proclaims.
10. "Dr. Clarke's second reply," *ibid.*, p. 41. *Intelligentia supramundana,* or more exactly, *extra mundana,* is an expression of Leibniz; cf. *Théodicée,* §217.
11. *Ibid.*, p. 45.
12. "Mr. Leibniz's third paper," *ibid.*, p. 57.
13. *Ibid.*, p. 59.
14. *Ibid.*, p. 69.
15. "Dr. Clarke's third reply," *ibid.*, p. 77. Dr. Clarke uses the term "property" in his own "replies" as well as in the translation of Leibniz's "papers" — and one understands full well why he does not use the more correct one, "attribute": just because Leibniz has mentioned Spinoza. But Leibniz himself *uses* the term "attribute"; moreover the French translation of Clarke's "replies," reviewed and acknowledged by Clarke himself, uses "attribute" for "property."
16. Dr. Clarke's example is rather bad as, in this case, there would be a *relative* displacement of "our world" in respect to the fixed stars.
17. The use of the principle of inertia in the discussion of the old problem whether God can move the world in a straight line (cf. my paper quoted *supra*, cap. III, n. 43) is rather ingenious.
18. "Dr. Clarke's third reply," *ibid.*, p. 85.
19. For Leibniz reality and individuality are inseparable.
20. "Mr. Leibniz's fourth paper," *ibid.*, p. 97.
21. *Ibid.*, p. 103.
22. Thus, practically, Leibniz and Descartes are in full agreement.
23. "Mr. Leibniz's fourth paper," *ibid.*, pp. 115 sq.
24. *Ibidem.*
25. *Ibidem.* Leibniz will mention Henry More in his fifth paper, n. 48: "To conclude. If the space (which the author fancies) void of bodies is not altogether empty: what is it then full of? Is it full of extended spirits perhaps, or immaterial susbtances, capable of extending and contracting of themselves; which move therein and penetrate each other without any inconveniency, as the shadows of two bodies penetrate one

another upon the surface of a wall? Methinks I see the revival of the odd imaginations of Dr. Henry More (otherwise a learned and well meaning man) and of some others who fancied that those spirits can make themselves impenetrable whenever they please."

26. *Ibidem.*
27. *Ibidem.*
28. *Ibid.*, p. 101.
29. *Ibidem.*
30. "Dr. Clarke's fourth reply," *ibid.*, p. 121.
31. We even have to suppose it if we want to link atomism with mathematical philosophy.
32. *Ibid.*, p. 125.
33. *Ibidem.*
34. *Ibidem.*
35. *Ibid.*, p. 127.
36. *Ibid.*, p. 131.
37. It is rather interesting to see Dr. Clarke use Henry More's famous concept and term.
38. *Ibid.*, p. 127.
39. *Ibid.*, p. 135.
40. *Ibid.*, p. 139.
41. *Ibid.*, p. 139.
42. *Ibid.*, p. 141.
43. *Ibid.*, p. 149.
44. *Ibid.*, p. 151.
45. This latter behaviour is, more often than not, branded as "arbitrariness."
46. "Mr. Leibniz's fifth paper," *ibid.*, p. 181.
47. *Ibidem.*
48. *Ibid.*, p. 211.
49. *Ibid.*, p. 183.
50. *Ibid.*, p. 207.
51. *Ibid.*, p. 231.
52. *Ibid.*, p. 189.
53. *Ibid.*, p. 193.
54. *Ibid.*, p. 195.
55. *Ibidem.*
56. *Ibid.*, p. 235.
57. *Ibid.*, p. 259.
58. *Ibid.*, pp. 269 sq.
59. "Dr. Clarke's fifth reply," *ibid.*, p. 295.

60. *Ibid.*, p. 313.
61. *Ibid.*, pp. 301 sq.
62. *Ibid.*, p. 349.
63. *Ibid.*, p. 367.
64. *Ibid.*, p. 335.

Index

Printed in the United States
94461LV00001B/16/A